The biology of human ageing

Edited by

A.H. BITTLES

Department of Anatomy and Human Biology, King's College, University of London

K.J. COLLINS

St Pancras Hospital, University College Hospital School of Medicine, London

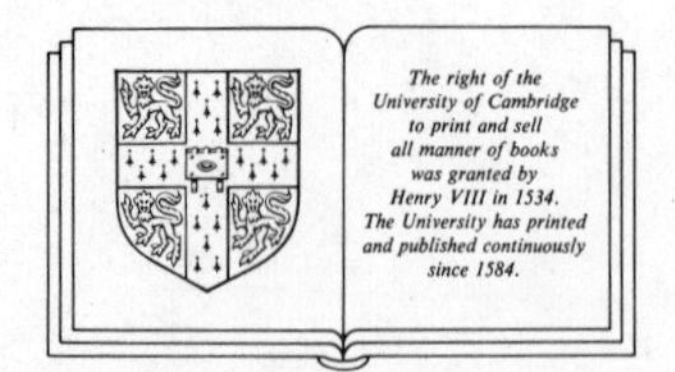

CAMBRIDGE UNIVERSITY PRESS

Cambridge

London New York New Rochelle

Melbourne Sydney

Published by the Press Syndicate of the University of Cambridge
The Pitt Building, Trumpington Street, Cambridge CB2 1RP
32 East 57th Street, New York, NY 10022, USA
10 Stamford Road, Oakleigh, Melbourne 3166, Australia

First published 1986

Printed in Great Britain at the University Press, Cambridge

British Library Cataloguing in Publication Data
The Biology of human ageing.
1. Ageing
I. Bittles, A.H. II. Collins, K.J.
612′.67 QP86

Library of Congress cataloguing in publication, data available

ISBN 0 521 30485 7

CONTENTS

PREFACE

During the last decade there has been a marked upsurge of interest in the topic of human ageing. This is largely due to the realization that the elderly comprise an ever-increasing proportion of the population and, at least in industrialised nations, age-related disorders now account for a substantial component of the expenditure on health care. In recent years, trends in ageing research have appeared to focus either at the cellular and sub-cellular levels or on specific topics in geriatric medicine. The aim of a two-day symposium held at Chelsea College, University of London, in April 1984, organised on behalf of the Society for the Study of Human Biology and the British Society for Research on Ageing, was to help bridge these two specialist areas by considering human ageing within the interdisciplinary context of human biology. In the proceedings presented in this book, the editors (also the organisers of the symposium) have brought together scientific papers presented by a group of distinguished research workers in gerontology and geriatrics. Although it obviously was not possible to achieve a fully comprehensive treatment of the subject, many of the underlying principles and current views on human senescence are considered, including stochastic and non-stochastic theories of ageing, biological markers in age assessment, and demographic, social and clinical features of ageing populations.

To biologists of whatever persuasion, it is evident that a phenomenon which is common to all species must be associated with, if not an integral part of, the evolutionary process. Accordingly, the initial chapter is concerned with a model to explain ageing in terms of energy investment and natural selection. In the following three chapters critical assessments are made of the efficacy of model systems, including specific insect mutant strains and human cell cultures, in testing theories of ageing, while age-related changes in the transmission of information are analysed at gene level. Since many workers consider ageing as essentially a developmental process, a chapter on the estimation of maturity during pubertal and pre-pubertal stages is included. Developing the theme, the marked effects exerted on biological age scores by social factors such as educational background and marital status are illustrated in the data from a U.S.-based longitudinal study and, by way of contrast, the inherent problems in attempting to characterise palaeodemographic structure from skeletal records are presented for a Romano-British population. An alternative aspect of ageing, dealing with differential resource partition between and within organs as part of the pattern of senescence, is then examined. The third major theme -

demography - is introduced by projections relating to the scale of increase in the proportion and actual numbers of those over retirement age in U.K. and U.S. populations. There is some divergence of opinion as to the effects of these changing age profiles, but the patterns of change in the two countries are similar and appear established. This is not the case in the U.S.S.R. which traditionally has claimed large numbers of centenarians. The chapter analysing and effectively rebutting such claims may at least partially explain recent, somewhat surprising Russian predictions of no increase or even an overall decline in life expectancy. In view of the conflicting predictions presented in the studies relating to the 'epidemiology' of ageing, the last chapter in this section on the health of an ageing population provides a valuable insight into the probable health needs of the elderly during the next decades. Selected topics on the physiology and biochemistry of ageing comprise the final four chapters of the book. While characteristic changes can be demonstrated in a number of biochemical parameters with age, the interaction of many causative factors makes interpretation difficult. Although the mechanisms of dietary restriction in extending mammalian lifespan also remain elusive, detailed information on the impact of dietary manipulation in ageing animals is provided in a comprehensive review of the subject. In the two final chapters on physiological aspects of ageing, physical activity patterns in the elderly before and after retirement and the effects of ageing on homeostasis, especially body temperature regulation, are discussed. Thus, throughout the book, human ageing is considered in terms of its biological, behavioural and clinical significance and it is this integrative approach which makes the contributions of particular relevance to the modern study of the subject.

It is a pleasure to thank the many individuals who helped with the Symposium and in particular the Chairmen of the four main sessions, Professor Derek Roberts, Mr. Clive Turner, Professor Norman Exton-Smith and Professor John Phillips, and our colleague Dr. Roger Briggs, Secretary of the British Society for Research on Ageing. We are grateful to the authors and journals who have given permission for the reproduction of figures and illustrations in the book, acknowledgements for which appear in the text. Additional financial support for the Symposium was provided by the Royal Society and by Farmitalia Carlo Erba, Imperial Chemical Industries, Wyeth, and Sandoz Products, to all of whom we wish to express our thanks.

A.H.B.

K.J.C.

AGEING AS A CONSEQUENCE OF NATURAL SELECTION

T.B.L. KIRKWOOD and R. HOLLIDAY
National Institute for Medical Research,
The Ridgeway, Mill Hill, London, U.K.

INTRODUCTION

The process of ageing is familiar in all human societies. From early in childhood we learn to recognise the signs of senescence, and quickly we become adept at combining a variety of visual and behavioural clues to derive a reasonably accurate impression of a person's age. This awareness of intrinsic mortality extends also to domestic and captive animals, and it leads almost naturally to the view that ageing is a form of inescapable deterioration which afflicts complex organisms in much the same way that mechanical breakdown occurs in machines. In consequence, we may be all too ready to take ageing for granted and to overlook the important question of why ageing should occur at all.

There are two reasons, at least, why we should be cautious about accepting the mechanistic view that ageing is inevitable. Firstly, ageing is not intrinsic to all living organisms. Bacteria and many eukaryotic microorganisms can divide indefinitely, many plants are capable of unlimited vegetative propagation, and some simple animals, such as coelenterates, have regenerative powers which apparently allow them to escape senescent changes (Comfort, 1979). Secondly, as Williams (1957) observed, the remarkable thing about ageing is that after its tremendous feat of morphogenesis a complex organism should be incapable of the seemingly much simpler task of merely maintaining what is already formed. Damage arises continually in organisms and their cells and powerful mechanisms exist to correct or repair these defects. There is no reason in principle why these systems should not be able to maintain the organism as a whole in a physiological steady state, as must be the case for the immortal organisms mentioned above and also for the germ line of all sexually reproducing species.

As early as the latter half of the last century, it was recognised that ageing and the duration of life were characters that required explanation in

terms of natural selection (Weismann, 1891; see, also, Kirkwood & Cremer, 1982). Since then, numerous authors have considered ageing from an evolutionary point of view. In this paper we review briefly the main general theories on the evolution of ageing and we develop further a 'disposable soma' theory which we have described previously (Kirkwood, 1977; Kirkwood & Holliday, 1979; Kirkwood, 1981). We consider also some of the special features of ageing in humans and discuss how these may have evolved.

DEFINITION OF AGEING

There are two quite distinct ways to define the ageing process. The first is through the physical changes that take place in individuals, which are manifested as a decline in a variety of body functions. The difficulty with this definition is that although the general features of ageing are similar from one individual to the next, the specific details vary quite considerably (see Borkan, this volume). Some go grey while others do not, some develop senile cataracts while others do not, and so on. The second definition, which is more useful from an evolutionary point of view, concerns the net effect of all these changes on the ability of the organism to survive. This is measured not for an individual, but for a population.

At the population level, the most concise definition of ageing is that the overall process of progressive, generalised impairment of the functions of organs and tissues results in an increasing age-specific death rate. This is illustrated in Figure 1 with a survival curve for a typical human population enjoying privileged conditions of social and medical care. Similar survival curves are found also for domestic and laboratory animals reared under conditions where mortality due to environmental hazards is kept to a minimum (Comfort, 1979). These curves are characterised by having reasonably high survival for the earlier part of the lifespan, followed by a steady decline thereafter until survivorship reaches zero at a maximum age which is a constant for the species.

Gompertz (1825) noted that for humans the form of the age-specific mortality rate curve is closely approximated by a simple exponential, and Sacher and Staffeld (1972) showed more recently that the same form of relationship holds also for several other species of mammals. A convenient representation of this pattern is given by the Gompertz-Makeham equation

$$\mu_x = \mu_o e^{\beta x} + \gamma$$

where μ_x is age-specific mortality at age x and the parameters μ_o, β and γ can

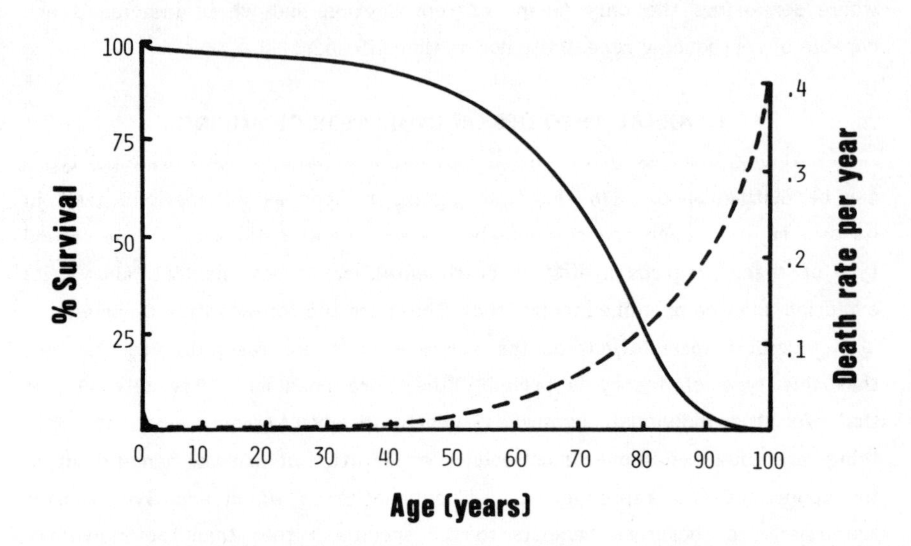

Figure 1: Survivorship (continuous curve) and mortality rate (dashed curve) as functions of age for a typical industrialised human society.

be termed 'basal vulnerability', 'actuarial ageing rate' and 'environmental mortality', respectively (see Sacher, 1978; Kirkwood, 1985). Note that although the Gompertz-Makeham equation gives μ_o as the mortality rate at age zero, the true infant mortality rate is usually greater than this and the equation should not be assumed to hold outside the adult age range over which it is normally fitted. The advantage of adding the third parameter, γ, to the equation to represent age-independent mortality due to the environment is that by keeping the intrinsic parameters μ_o and β fixed, but allowing γ to vary, the same equation can be used for both wild and protected populations of the same species (see Kirkwood, 1985).

The definition of ageing in terms of a survivorship pattern showing a Gompertzian increase in mortality with age is strictly meaningful only for a species where the individual organism is clearly defined and where the act of reproduction does not in itself signal the end of the lifespan. For species capable of vegetative reproduction and for semelparous species which reproduce once only and die soon afterwards, usually in a highly determinate fashion, the definition of ageing is more complex (see Kirkwood & Cremer,

1982; Kirkwood, 1985). In this paper, we confine attention to higher animals where sex offers the only means of reproduction and where individuals are capable of reproducing repeatedly during their lifespan.

GENERAL THEORIES ON EVOLUTION OF AGEING

Among theories to explain the evolution of ageing, two principal types can be distinguished. In the first, ageing is seen as a beneficial trait in its own right. Such theories can be termed adaptive theories. The second type of theory regards ageing as detrimental, or at best neutral, and so its evolution must be explained indirectly. These are the non-adaptive theories.

Explicit formulations of the adaptive view are rare, despite the fact that this type of theory is probably the more popular. The difficulty is that, for the individual, ageing is clearly disadvantageous, since if other things are equal senescence must reduce expectation of life and hence diminish the opportunity for reproduction. The benefits on which adaptive theories are based are therefore benefits to the species, rather than the individual. One idea, which can be traced to Weismann's (1891) original writings, is that without ageing a species might overcrowd its environment and exhaust its resources (see also Wynne-Edwards, 1962). A second is that ageing accelerates the turnover of generations and thereby increases the chance of a species adapting to a change in its environment (Woolhouse, 1967; von Weizsacker, 1980). The flaw with both these arguments, as with the adaptive theory in general, is that on the one hand ageing is so rarely encountered in organisms in the wild that the need and opportunity for an adaptive ageing process to have evolved would have been minimal (Medawar, 1952), while on the other, group selection for advantage to the species is so much weaker than selection for advantage to the individual (Maynard Smith, 1976) that any adaptive mechanism for ageing would be selectively unstable. In a group of animals which age, individuals which did not do so would be at a selective advantage.

Non-adaptive theories on the evolution of ageing belong to two sub-types, the first suggesting that natural selection is simply unable to prevent the deterioration of older organisms because it becomes attenuated with age, while the second suggests that ageing is a by-product of selection for other beneficial traits. Both sub-types share a common principle originally pointed out by Haldane (1941), and later clearly enunciated by Medawar (1952), which asserts that the force of natural selection reduces progressively

with age. The reason for this is that selection on genes acting early in life will affect a greater proportion of individuals than genes acting late, when the proportion of survivors will be smaller and the remaining fraction of their lifetime expectation of reproduction will be less (see, also, Williams, 1957; Hamilton, 1966; Edney & Gill, 1968; Kirkwood & Holliday, 1979; Charlesworth, 1980).

Medawar (1952) proposed that the attenuation with age of the force of natural selection was sufficient by itself to explain the evolution of ageing. He suggested that selection acting on genes with age-specific times of expression would tend to defer the age of expression of harmful genes, so as to minimise their potential for deleterious effects. Once a harmful gene had been so far deferred that it was expressed at an age when survivorship in the wild environment was effectively zero, natural selection could no longer act to further postpone its expression, nor could it act to eliminate the gene altogether. Over time, there might thus accumulate a 'genetic dustbin' of late-acting deleterious genes which in the normal environment would not have opportunity to be expressed, but which in a protected environment would combine to handicap severely any individual which by virtue of reduced environmental mortality lived long enough to encounter their detrimental effects. This process, Medawar (1952) suggested, could account for the evolutionary origin of senescence.

That Medawar's hypothesis has relevance to the evolution of ageing seems highly probable, but that it offers a sufficient explanation for its origin seems less clear. Experimental studies in *Drosophila* by Rose and Charlesworth (1980) failed to detect any increase in additive genetic variance with age in female fecundity, contrary to expectation if late-acting mutations were accumulating, while the theoretical objection has been raised that in the absence of ageing it is hard to see what would be the timing mechanism to determine 'lateness' in a life-history that potentially could last indefinitely (Kirkwood, 1977; Sacher, 1978; Kirkwood & Holliday, 1979).

The second sub-type of non-adaptive theory, that ageing is a by-product of selection for other beneficial traits, was formulated in general terms by Williams (1957). Williams' theory was derived from an argument similar to Medawar's except for the crucial difference that the genes in question were assumed to be *pleiotropic*, the same genes having both good effects early and bad effects late. Natural selection, it was argued, should favour retention of the genes on the basis of their benefits, but defer as far as possible the time of

expression of the deleterious effects to ages when survivorship would in any case be low. The diminution of the force of natural selection with age would ensure that even quite modest early benefits would outweigh severe harmful side-effects, provided the latter occurred late enough. The result, as with Medawar's theory, is that genes with late-acting deleterious effects would accumulate and give rise to senescence in any individuals which lived long enough. In Williams' theory, ageing was attributed to the positive action of selection on the pleiotropic genes, but was not *in itself* regarded as beneficial. The theory was, therefore, genuinely a non-adaptive one.

Williams' theory avoids the objections raised against Medawar's, since (i) it is easier to see how selection on a pleiotropic gene expressed early in life may modulate the age at which deleterious effects become apparent and (ii) the theory does not necessarily imply an increase in genetic variance with age. The theory does not, however, suggest which particular kinds of pleiotropic gene are likely to be involved in ageing and therefore its predictions are of a very general nature. In the remainder of this paper, we describe a more specific version of the idea that ageing is a by-product of selection for other benefits, which leads to more definite predictions. This is the view we have termed the disposable soma theory.

DISPOSABLE SOMA THEORY

The starting point for the disposable soma theory is to consider the organism as a 'black box' which takes up energy from the environment, in the form of nutrients and other resources, and converts it into progeny. The law of natural selection states, effectively, that the genotypes which are the most efficient in this process are the ones most likely to survive. Of the energy it takes in, an organism must allocate some to growth, some to defending itself against predators and disease, some to reproduction, and so on. In the present context, our particular interest is in the relative allocation of resources to maintenance and repair of the non-reproductive bodily tissue, or soma, and to reproduction. Maintenance and repair is taken to include the prevention and removal of DNA damage, accuracy in macromolecular synthesis and the degradation of defective proteins, wound healing, and the immune response. Clearly, the more of its energy the organism allocates to somatic maintenance the less is available for reproduction, and conversely. The balance which must be struck is between the competing benefits of (i) living longer by being better able to cope with random damage, and thereby being able to reproduce over a

greater timespan, or (ii) reproducing at a greater rate.

To be specific about how the organism should optimally allocate its resources between somatic maintenance and reproduction, we require some measure of fitness by which the success of alternative strategies may be judged. The most convenient measure for this purpose, and the one that will be adopted here, is the *intrinsic rate of natural increase*, r, in the population, r being defined as the unique root of the equation

$$\int_0^\infty e^{-rx}\, l(x)\, m(x)\, dx = 1$$

where l(x) denotes survivorship at age x (expressed as a fraction) and m(x) denotes reproductive rate at age x. Both l(x) and m(x), and hence r, will be affected by varying the investment in somatic maintenance and repair, and the question is whether there is some optimum level of investment for which r is a maximum.

Intuitively, it is clear that an optimum level of somatic maintenance and repair does exist. If too many resources are invested in maintenance, too few will be available for other vital activities, especially reproduction. On the other hand, if too few resources are invested in maintenance, the soma will disintegrate from accumulated random damage at an early age, possibly even before it becomes reproductively mature.

The second vital question is whether the optimum level of investment in somatic maintenance and repair is greater or less than the minimum level at which the organism can keep the rate of occurrence of damage low enough that it can be repaired at the same rate that it arises. Above this minimum level the organism will remain in a steady state, but below it damage will accumulate progressively. That such a threshold may, in fact, exist was implied by the premise with which we introduced this paper, namely that progressive wear and tear is not inevitable. The disposable soma theory is derived from the conclusion that the optimum level of investment in somatic maintenance and repair is always less than the level required for indefinite somatic survival. We justify this conclusion as follows.

No species is immune from purely random mortality exacted by the environment. There is therefore no advantage to be gained from investing in potential somatic immortality when in practice the return from this investment cannot be realised. For example, suppose there exists a species which invests sufficient resources in somatic maintenance and repair to be potentially capable of living forever, but which is subject to environmental mortality of 10% per

year. Individuals at 100 years of age would be physiologically as young as individuals of, say, 10 years, but the chance of surviving to be 100 is about 0.000027. For practical purposes this chance is negligible, and in consequence it would clearly be preferable to reduce the investment in maintenance so somatic viability is preserved only up to, say, 50 years of age and to use the extra resources liberated by this saving to increase reproduction. The upshot is that natural selection will favour the strategy of investing only sufficient resources in somatic maintenance and repair to ensure that the soma remains viable through its normal expectation of life in the wild, and for individuals protected from the usual mortality of the natural environment ageing will occur as accumulated damage to the soma eventually takes its toll.

To justify the disposable soma theory more formally will require that the dependence of $l(x)$ and $m(x)$ on the level of investment in somatic maintenance and repair be formulated explicitly. At present, the precise physiological trade-offs are unknown, so this can be done only in the context of specific mathematical models. We illustrate a particular instance of such a model below, and a more general treatment will be given elsewhere (Kirkwood, in preparation).

As regards survivorship, $l(x)$, we assume a Gompertzian mortality function for which the parameters affected by repair are basal vulnerability, μ_o, and actuarial ageing rate, β. We define a variable ρ to represent the fractional investment in somatic maintenance and repair, $\rho = 0$ corresponding to zero repair and $\rho = 1$ corresponding to the maximum which is physiologically possible. We also assume that beyond some fixed level of repair ρ' $(0 < \rho' < 1)$ damage does not accumulate and the organism is potentially immortal. μ_o and β are both decreasing functions of ρ and we suppose

$$\mu_o = \mu_{min}/\rho$$

$$\beta = \begin{cases} \beta_o(\rho'/\rho - 1) & \rho < \rho' \\ 0 & \rho > \rho' \end{cases}$$

The assumption $\beta = 0$ for $\rho > \rho'$ corresponds with our definition of ρ' as the repair level beyond which ageing does not occur, although there will still be reduction in μ_o. We assume neonatal survivorship is proportional to ρ, with Gompertzian mortality acting thereafter. These assumptions give

$$l(x) = \rho e^{-(e^{\beta x}-1)\mu_o/\beta - \gamma x}$$

where the dependence of μ_o and β on ρ is as defined above.

For reproductive rate, m(x), the primary parameters affected by ρ are assumed to be age at maturation, a, which will be increased by raising ρ, and initial reproductive rate, f, which will be a decreasing function of ρ. We assume

$$a = a_0/(1-\rho)$$

$$f = f_{max}(1-\rho)$$

and we further suppose that for $\beta > 0$, i.e. for a population with a non-zero ageing rate, the reproductive rate declines due to senescence with the same

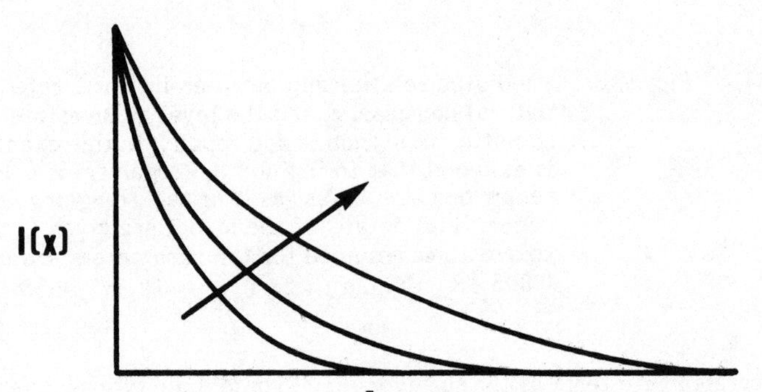

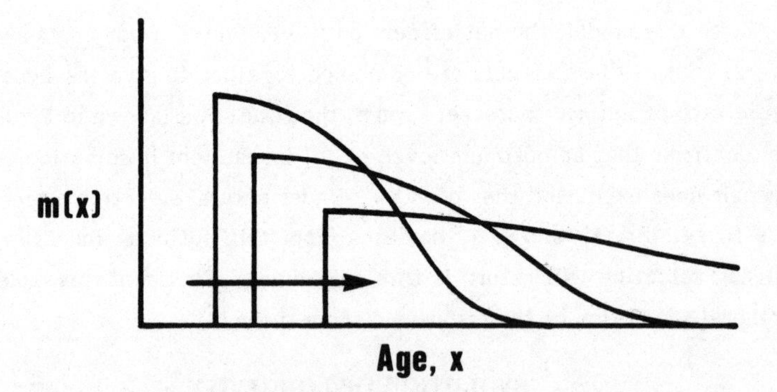

Figure 2. Effects on survivorship, l(x), and reproductive rate, m(x), of varying the investment, ρ, in somatic maintenance and repair. The arrows indicate the direction of increasing ρ. For full details of the model, see text.

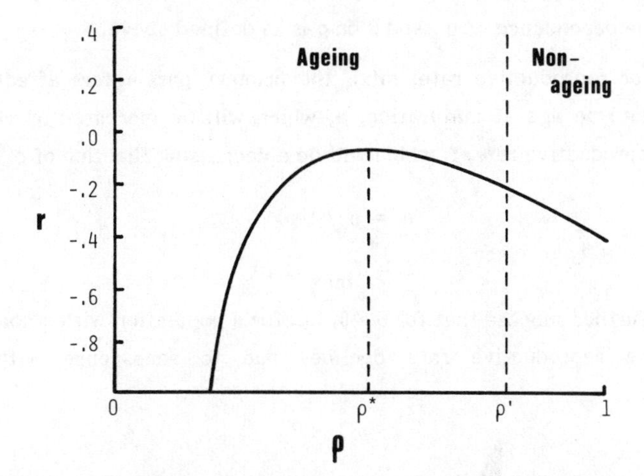

Figure 3. Predicted relationship between intrinsic rate of natural increase, r, and the level of investment, ρ, in somatic maintenance and repair. In this example, it is assumed that for $\rho > 0.8$ the organism is able to repair damage as fast as it arises, so ageing does not occur. Full details of the model are given in the text; values assumed for the parameters are $\mu_{min} = 0.005$, $\beta_o = 0.5$, $a_o = 2$, $f_{max} = 10$, $\gamma = 0.38$.

Gompertzian rate term as for survivorship. These assumptions give

$$m(x) = fe^{-(e^{\beta x}-e^{\beta a})\mu_o/\beta} \qquad x \geqslant a$$

With this model, the net effects on l(x) and m(x) of varying ρ are shown in Figure 2. When these effects are combined together to give the dependence of intrinsic rate of natural increase, r, on ρ, the result is as shown in Figure 3. The model confirms that an optimum level, ρ*, of investment in somatic maintenance and repair does exist, and that $\rho^* < \rho'$. Under natural selection a species should evolve to set $\rho = \rho^*$ and if it deviates from this optimum in either direction stabilising selection will return it to equilibrium. This confirms quantitatively the general prediction of the disposable soma theory.

EVOLUTION OF LONGEVITY

For any theory on the evolution of ageing, an important secondary problem is to explain how species have evolved to have different lifespans. Clearly, longevity is closely linked with other life-history attributes, since a species which produces few progeny at long intervals will, in general, need to live longer than one which has large, frequent litters. For the disposable soma

theory the question is how the optimum investment in somatic maintenance and repair may be expected to vary from one ecological niche to another.

One of the main factors to be taken into account in optimising a species' life-history is the level of environmental mortality to which it is exposed. This may be represented through the parameter γ of the Gompertz-Makeham equation. For large γ, i.e. for high environmental mortality, the optimum life-history is likely to place greater emphasis on a high reproductive rate and lesser emphasis on investment in somatic maintenance and repair, since in this type of environment the individual's expectation of life is in any case severely limited. Major investment in somatic maintenance and repair would therefore be a waste. Conversely, a species subject to low environmental mortality may profit by investing relatively heavily in somatic maintenance and repair.

The model described in the previous section may be used also to investigate the effects on ρ^* of varying components of a species' life-history. As an example, we consider an adaptation, such as evolution of a relatively large brain, which brings about a reduction in environmental mortality, but which also causes reduction in the maximal reproductive rate, f_{max}. Such an adaptation requires initially that the lowering of environmental mortality raises the intrinsic rate of natural increase, r, despite the reduction in reproductive rate. Eventually, however, for a species in equilibrium with its environment the value of r must settle at zero. Introducing these variations into the model, subject to the constraint that for $\rho = \rho^*$ the intrinsic rate of natural increase is zero, the results given in Table 1 were obtained. These confirm the expected correlations between environmental mortality, reproductive rate and ρ^*, and they show

Table 1. Relationships between environmental mortality, reproductive rate at maturation (f), optimum level of investment in somatic maintenance and repair (ρ^*), and longevity, predicted for four hypothetical species according to the disposable soma theory. See text for details of the model.

Environmental mortality (% per year)	Reproductive rate, f (per year)†	Optimum level of maintenance and repair, ρ^*	Species' longevity (years)≠
40	10.4	.46	8
20	3.2	.54	12
10	1.0	.65	28
5	0.4	.74	57

† births (both sexes) per female
≠ 99th percentile

corresponding correlations with longevity. Although the predicted correlations of environmental mortality, reproductive rate and mortality are not in themselves novel, the key contribution of this analysis is to establish a mechanism, namely varying the investment in somatic maintenance and repair, through which they may have evolved. The disposable soma theory can thus account for the divergence of species' lifespans, as well as for the evolution of ageing itself, and specifically it predicts that direct correlations will be found between species' longevities and their levels of efficiency in somatic maintenance and repair.

SPECIAL FEATURES OF HUMAN AGEING

Two striking features distinguish ageing in human populations from ageing in other animals. Firstly, as noted in the Introduction, ageing is a commonplace occurrence in human society and has apparently been familiar for at least several thousand years. This is in sharp contrast with other species, for which ageing is either not seen in the wild or is extremely rare. Secondly, in human females reproduction is brought to an abrupt close by the menopause, usually in the fourth or fifth decade of life. In other species, except those which reproduce once only anyway, reproduction continues throughout life, albeit at an increasingly irregular and declining rate. (It has been suggested that female chimpanzees and macaques exhibit a partial form of menopause (Graham et al, 1979; Gould et al, 1981), although this is not sharply defined.)

The frequent occurrence of senescence in humans and the apparently programmed cessation of female reproduction both appear to challenge the superiority of non-adaptive theories of ageing over adaptive ones. As ageing is best known to us through observation within our own species, this may partly explain why the adaptive view retains such popularity. In fact, however, our species is such a newcomer to the evolutionary scene that it makes little sense to construct general theories on the evolution of ageing based on features unique to humans. More plausibly, the special aspects of human ageing are to be explained through secondary modification of the general non-adaptive theory outlined above.

The most salient respect in which humans differ from other species is in the size of the brain relative to that of the body. This makes possible our complex social behaviour, which affords significant protection against environmental hazard, but at the cost of greatly delaying maturation and prolonging the dependence of children on their parents. These latter factors necessitate considerable longevity and we have shown already how this may have come about

through selection for increased somatic maintenance and repair. There is, however, a limit to the extent that resources can be spared for increased somatic maintenance when reproductive maturity is already long delayed and when further delay will seriously lower the intrinsic rate of natural increase, r. Further application of the model described earlier suggests there may come a point where extra investment in somatic maintenance and repair is no longer beneficial, even though environmental mortality may have fallen to a level where senescence now acts to limit survivorship. At this stage, ageing ceases to be merely a latent possibility, affecting only the rare individual which survives environmental hazards, and becomes a general feature in the population.

Once ageing is common, evolution of the menopause can be expected to follow as a secondary adaptation to the enhanced risk of attempting to continue reproduction with an ageing soma. For the human female, reproduction is a physically demanding process made more difficult by the large brain size of the neonate. To sustain fertility throughout life would greatly increase the likelihood that death would occur as a consequence of pregnancy. This, together with the dependence of existing young on continued parental care and protection, makes it highly plausible that the menopause can be explained as a mechanism to limit female reproduction to ages when it is comparatively safe.

In this way, both of the special features of human ageing can be accounted for quite readily within the non-adaptive framework of the disposable soma theory.

CONCLUSIONS

Arguments to account for the evolution of ageing as an adaptation in its own right are generally unsound. It is therefore from among the non-adaptive theories that the explanation must be sought. The mainstay of the non-adaptive theories is the principle that even in the absence of senescence the losses caused by environmental mortality mean that events occurring at later ages have lesser evolutionary significance (Medawar, 1952; Williams, 1957).

The non-adaptive theory which offers the most specific mechanism to account for the general evolution of ageing is the disposable soma theory (Kirkwood, 1977; Kirkwood & Holliday, 1979; Kirkwood, 1981). This suggests that ageing has evolved as a by-product of optimising the organism's allocation of resources among the various tasks it must perform. In particular, the disposable soma theory proposes that the optimum investment of resources

in somatic maintenance and repair is always less than the minimum which would be required for indefinite somatic survival. The consequence is that ageing results through the progressive accumulation of somatic defects and damage, and the theory further accounts for the divergence of species' lifespans through the differential tuning of the optimum level of somatic maintenance and repair to take account of the various constraints imposed by different ecological niches.

A direct means to test the predictions of the disposable soma theory, and in the process possibly to illuminate the primary causes of ageing, is provided by comparative study of the efficiency of different systems for maintenance and repair of the soma. No single process is likely to account for ageing in all the diverse forms of organisms in which it occurs, and some discretion is likely to pay dividends in selecting primary targets for research. Among mammals it appears there may be an intrinsic ageing process limiting the capacity of cell lineages to proliferate (see Bittles & Sambuy, this volume), so there is reason to focus initially on mechanisms for intracellular maintenance and repair. The most fundamental of intracellular maintenance mechanisms are those responsible for the correction or elimination of errors made during the synthesis of macromolecules, and elsewhere we have suggested that these systems deserve special attention (Kirkwood & Holliday, 1979; Holliday, 1984). Should errors in macromolecules prove, however, not to be the primary cause of cellular ageing, investigation would proceed naturally to higher levels of cellular organisation. At present, the technical problems of assessing the contribution of errors in the synthesis of macromolecules to the loss of viability of cells remain formidable (Kirkwood et al, 1984).

The ageing of humans is of special interest, firstly because it concerns all of us directly, and secondly because humans are unique in the extent to which ageing is seen "in the wild" and in possession of a clear cut menopause. The disposable soma theory suggests both these features may be explained as secondary modifications to the basic process through which ageing and longevity have evolved. It should be noted, however, that our species is not yet at equilibrium, as witnessed by our continuing population growth, so the genetic determinants of our ageing process may not yet have been optimised. An understanding of how these determinants are shaped could, therefore, be of significance in helping us to predict how, in the long term, our life history may be influenced through the further action of natural selection.

REFERENCES

Bittles, A.H. & Sambuy, Y. (1985). Cell culture systems in the investigation of human ageing. (This volume.)

Borkan, G.A. (1985). Biological age assessment in adults. (This volume.)

Charlesworth, B. (1980). Evolution in Age-structured Populations. Cambridge: Cambridge University Press.

Comfort, A. (1979). The Biology of Senescence, 3rd edn. Edinburgh: Churchill Livingstone.

Edney, E.B. & Gill, R.W. (1968). Evolution of senescence and specific longevity. Nature, **220**, 281-282.

Gompertz, B. (1825). On the nature of the function expressive of the law of human mortality and on a new mode of determining life contingencies. Philosophical Transactions of the Royal Society, **2**, 513-585.

Gould, K.G., Flint, M. & Graham, C.E. (1981). Chimpanzee reproductive senescence: a possible model for the evolution of the menopause. Maturitas, **3**, 157-166.

Graham, C.E., Kling, O.R. & Steiner, R.A. (1979). Reproductive senescence in female nonhuman primates. In: D. M. Bowden (ed.), Aging in Non-Human Primates, pp. 183-202. New York: Van Nostrand Reinhold.

Haldane, J.B.S. (1941). New Paths in Genetics. London: Allen & Unwin.

Hamilton, W.D. (1966). The moulding of senescence by natural selection. Journal of Theoretical Biology, **12**, 12-45.

Holliday, R. (1984). The unsolved problem of cellular ageing. In: Monographs in Developmental Biology, volume 17, pp. 60-77. Basel: Karger.

Kirkwood, T.B.L. (1977). Evolution of ageing. Nature, 270, 301-304.

Kirkwood, T.B.L. (1981). Repair and its evolution: survival versus reproduction. In: C. R. Townsend & P. Calow (eds.), Physiological Ecology: an Evolutionary Approach to Resource Use, pp. 165-189. Oxford: Blackwell Scientific Publications.

Kirkwood, T.B.L. (1985). Comparative and evolutionary aspects of longevity. In: C. E. Finch & E. L. Schneider, Handbook of the Biology of Aging, pp. 27-44. New York: Van Nostrand Reinhold.

Kirkwood, T.B.L. & Cremer, T. (1982). Cytogerontology since 1881: a reappraisal of August Weismann and a review of modern progress. Human Genetics, **60**, 101-121.

Kirkwood, T.B.L. & Holliday, R. (1979). The evolution of ageing and longevity. Proceedings of the Royal Society, B**205**, 531-546.

Kirkwood, T.B.L., Holliday, R. & Rosenberger, R.F. (1984). Stability of the cellular translation process. International Review of Cytology, **92**, 93-132.

Maynard Smith, J. (1976). Group selection. Quarterly Review of Biology, **51**, 277-283.

Medawar, P.B. (1952). An Unsolved Problem in Biology. London: H. K. Lewis. (Reprinted in P.B. Medawar, 1981, The Uniqueness of the Individual, New York: Dover.)

Rose, M.R. & Charlesworth, B. (1980). A test of evolutionary theories of senescence. Nature, **287**, 141-142.

Sacher, G.A. (1978). Evolution of longevity and survival characteristics in mammals. In: E. L. Schneider (ed.), The Genetics of Aging, pp. 151-167. New York: Plenum.

Sacher, G.A. & Staffeld, E. (1972). Life tables of seven species of laboratory reared rodents. Gerontologist, **12**, 39.

Weismann, A. (1891). Essays upon Heredity and Kindred Biological Problems, volume 1, 2nd edn. Oxford: Clarendon Press.

Weizacker, C.F. von (1980). Ageing as a process of evolution. In: Conference on Structural Pathology in DNA and the Biology of Ageing, edited by L. Schoeller, pp. 11-20, Bonn: Deutsche Forschungsgemeinschaft.

Williams, G.C. (1957). Pleiotropy, natural selection and the evolution of senescence. Evolution, **11**, 398-411.

Woolhouse, H.W. (1967). The nature of senescence in plants. In: H. W. Woolhouse, Aspects of the Biology of Ageing: Symposia of the Society for Experimental Biology, No. XXI, pp. 179-213. Cambridge: Cambridge University Press.

Wynne-Edwards, V.C. (1962). Animal Dispersion in relation to Social Behaviour. Edinburgh: Oliver & Boyd.

GENETIC INFORMATION IN AGEING CELLS

S. P. MODAK[1], D. D. DEOBAGKAR[1], G. LEUBA-GFELLER[2], C. GONET[3] & S. BASU-MODAK[1]

[1]Department of Zoology, University of Poona, Pune, India;
[2]Institute of Anatomy-CHUV, University of Lausanne, Lausanne, Switzerland;
[3]Institute of Zoology, Science II, University of Geneva, Geneva, Switzerland

INTRODUCTION

Fertilization of an egg and the subsequent mixing of maternal and paternal chromosomal complements result in restitution of the diploid state composed of paired allelic sets of genes except, possibly, in the case of the male XY combination. The chromosomal basis of heredity and modality of the parent-to-progeny transfer by Mendelian genetic principles is well established. We also know that the molecular basis of heredity lies in the sequence of trinucleotides in DNA which is, thus, the primary carrier of genetic information. In each species the amount of DNA is maintained at a constant level and the law of DNA constancy is believed to imply that all cells in a lineage contain genetic information which is both qualitatively and quantitatively identical. It is well established that over 90% of the eukaryotic genome is noncoding but the "raison-d'etre" for noncoding sequences remains totally obscure. The law of DNA constancy is violated in the case of rDNA amplification during ovogenesis, chromosome elimination from the somatic cell lineage in nematodes, polytenization of chromosomes in diptera larvae, and polyploidization in both animal and plant cells (for details see Davidson, 1976). Cytogenetic maps have been described for a number of species and it is believed that, similar to the germ cells, somatic cells maintain these with fidelity. Since the studies of Hozumi and Tonegawa (1976) on the location of IgG genes, evidence has been accumulating to suggest that the position of one or more genes need not be identical with reference to either each other or, with respect to the total cytogenetic map, even in cells derived from the same lineage. The best example defying the Mendelian principle of the transfer of heredity is the transposable element (McClintock, 1956). However, only scanty information is available from eukaryotes on the relative number and types of transposition events altering the 'normal' arrangement of genes at fixed positions (Finnegan et al, 1981;

Spradling & Rubin, 1981). Finally, the analysis of cytogenetic maps has yet to answer the question on the extent to which the expressibility of a gene depends on its position relative to neighbouring genes. There is evidence showing that a gene position effect is exhibited by integrated viral genes (Jahner et al, 1982) and by cloned genes such as beta globin (Constantini et al, 1984).

The classical experiments on parthenogenesis and interspecies hybrids have established that the presence of alleles, in addition to allele-comparability, are essential factors for the maintenance of cell lineages. At the other end of the spectrum, one finds that 'transformed' cells exhibit 'immortality' and these contain a widely heterogeneous (aneuploid) chromosomal complement. The qualitative as well as quantitative nature of the allelic interaction is not understood. It is possible that cellular proliferation and survival is controlled by a subset of alleles which, in the progeny of aneuploid or heteroploid transformed cell lines, may be represented disproportionately. The high survival- and growth-potentials of such cells clearly suggest that the law of DNA constancy is not necessarily relevant to these properties.

The mechanism of retrieval of genetic information involves transcription of DNA into RNA which is then translated giving rise to the primary phenotype. It is implicit in this scheme that the genotype to phenotype conversion takes place with the same efficiency and fidelity for all genes, provided that the informational integrity of the genome is maintained. A large number of physical and chemical agents damage DNA, thereby impairing phenotypic expression (Hart et al, 1979). In all cells there exists a number of enzyme systems which are able to repair the damaged sites in DNA thus restoring the genotype to its original state. According to the most prevalent hypothesis (Modak, 1972; Modak & Unger-Ullmann, 1980), there exists a causal relationship between the repair of DNA and genome integrity, and this idea has gathered considerable support (see, Hart et al, 1979; Modak & Unger-Ullmann, 1980; Modak & Chalkley, 1981). It now appears that the repair of DNA damage is dependent on three factors, namely, the damaged site, the specific repair enzyme system, and accessibility of the damaged site to the repair system(s). DNA bases may become methylated or demethylated, thereby affecting their sensitivity to DNA damaging agents such as nucleases (Arber, 1974). In prokaryotes, 6mA stimulates mismatch repair (Radman et al, 1980) while 5mC stimulates recombination repair (Korba & Hays, 1982). Similarly, the extent of methylation has been correlated with both gene integrity and functional state (Razin & Riggs, 1980). Thus, a detailed analysis of these

various parameters is necessary in order to elucidate the molecular mechanism(s) involved in the maintenance or impairment of genome integrity in ageing cells.

We have examined some of the variables related to the control of genome integrity in postmitotic differentiation of neurones and liver cells of mice at ages distributed throughout the adult lifespan. We find that during the last quarter of the lifespan, cortical neurones accumulate single strand breaks in DNA, and liver chromatin acquires increasing resistance to the attack by Micrococcal nuclease. We also find that DNA in 'old' liver contains a significantly smaller proportion of 5mC residues as compared to very 'young' liver-DNA.

MATERIALS AND METHODS

F_1 hybrid mice (C57Bl x A/J) aged 1, 3, 6, 11, 16, 18, 22, 24 and 28 months were sacrificed by decapitation. Cerebral cortices (area 17) were fixed (80% ethanol, on ice, 2-3°C) embedded in paraffin (52-53°C), sectioned (5 µ thick) and mounted on glass slides (Modak et al, 1969). One-half of the deparaffinised slides were treated in 0.01N HCl (15 min, on ice) to denature DNA *in situ*, and the remainder were left as 'undenatured' (Modak et al, 1969; Modak & Bollum, 1970, 1972). Incubation wells were attached over the sections and the enzyme reaction mixture prepared, containing 100 µCi ^{3}H-dGTP (sp.A. 13 Ci/mM, Radiochemical Centre, Amersham, U.K.), 1000 units terminal deoxynucleotidyl transferase (Boehringer-Mannheim), 1mM $MnCl_2$, 1 mM DTT, 20mM HEPES (pH 7.4), 0.01% BSA (Fraction V), and 8% glycerol in a total volume of 1.0 ml. Twentyfive µl of the reaction mixture was added to each reaction well and the sections incubated for 1h at 37°C. The reaction was terminated by successive washes in 0.15M NaCl, 5% TCA containing 1% PPi, 5% TCA alone and running water (Modak & Bollum, 1970, 1972). Incubation wells were detached and slides were coated by dipping in the liquid emulsion L-4 (Ilford) diluted with water 1:1. Autoradiographs were developed after 48h in the developer D-11 (Kodak), fixed, and nuclei were lightly stained with Meyer's haemalum. From autoradiographs, grains per nucleus were determined by examining 200 nuclei from layer 3 of the cerebral cortex neurones.

From the same mice used above, and an additional number of 33 month-old animals, livers were dissected, the gall bladders discarded, and cell nuclei isolated according to Appleby and Modak (1977). Nuclei were digested with Micrococcal nuclease (4 units/10^{6} nuclei) for different times (30sec to 120min) as described before (Appleby & Modak, 1977). Aliquots of deproteinised lysates

were examined by electrophoresis in 1.6% agarose as before (Appleby & Modak, 1977). After treatment with RNase (25μg/ml) for 30min, gels were stained in ethidium bromide (2μg/ml) for 20min and DNA was visualised on a transilluminator (254nm) and photographed. In these gels SV40 DNA-Hind III restriction fragments were used as MW standards. From densitometric scans of negatives, the MW of the DNA fragments were estimated graphically (Appleby & Modak, 1977).

From another set of aliquots, TCA-precipitable DNA was estimated by the micro-diphenylamine technique as described by McMaster and Modak (1977) and the data plotted as % acid-precipitable DNA against the duration of digestion with Micrococcal nuclease.

Eleven-day newborn mice and 25.5-month-old adults were used for the extraction of nuclear DNA from liver by the SDS-EDTA-Proteinase K hydrolysis and phenol chloroform procedure followed by chloroform : iso amyl alcohol (24:1). DNA was then treated with RNase A and re-extracted as above. DNA was dissolved in TE buffer (10mM Tris-HCl, pH 7.4, 1mM EDTA) and 1μl samples (40μg/ml) were spotted on to a BA 85 nitrocellulose paper. Filters were baked for 4-6h at 65°C. The biotin avidin peroxidase assay for the detection of 5mC in DNA was carried out as described by Achwal and Chandra (1982). For this, nitrocellulose paper was treated with the antibody raised against 5mC (4-6h, antibody concentration 180μg/10ml), washed extensively and then treated with biotinylated goat anti rabbit antibody. The 5mC-antibody-anti IgG was detected by a complex of avidin and biotinylated peroxidase (Vector Laboratories, Burtingame, UK). Nitrocellulose filters were washed and stained for peroxidase with 0.1% 3-3' diaminobenzidine tetrahydrochloride and 0.02% H_2O_2. The intensity of the stained spot correlates directly with the amount of 5mC present in the DNA (Achwal et al, 1984).

RESULTS

Detection, in situ, of free 3'-OH ends in neuronal DNA

Figure 1 shows autoradiographs of mouse cerebral cortex sections in the regions of layer 3 of the visual area. The acid-denaturation causes an increased amount of incorporation of ^{3}H dGMP catalysed by terminal transferase. As compared to the neurones of the 3-month-old mouse, the grain count per nucleus of the neurones from the 28-month-old mouse is significantly higher. The data on grains per nucleus are obtained as mean number of grains and plotted against the age of the mouse (Fig. 2). It is seen that the

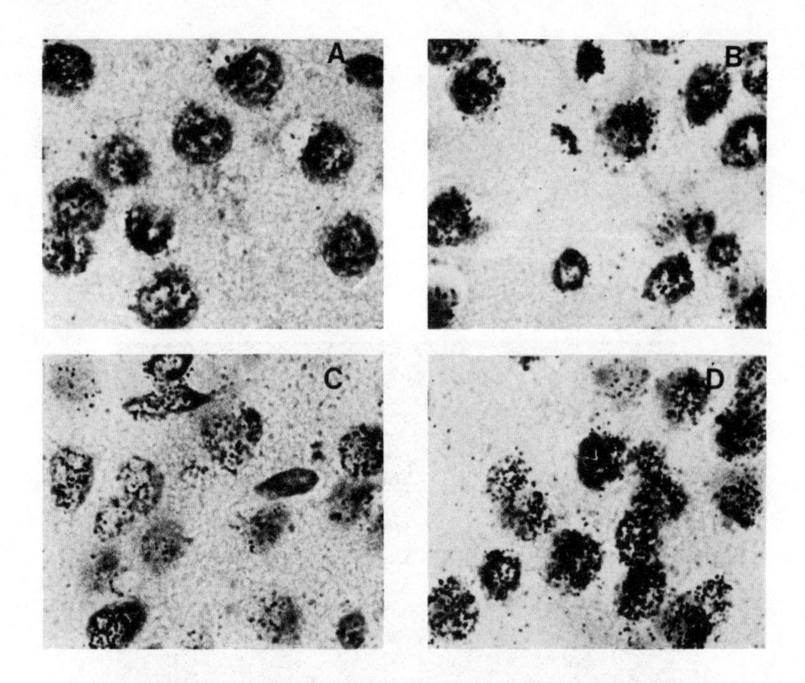

Figure 1. Autoradiographs of mouse cerebral cortical neurones from layer 3 of the visual area of sections incubated with terminal transferase and ^{3}H-dGTP. A: 3-month-old and 'undenatured'. B: 3-month-old and 'acid-denatured'. C: 28-month-old and 'undenatured'. D: 28-month-old and 'acid-denatured'. (Data derived from Modak et al, 1985a.)

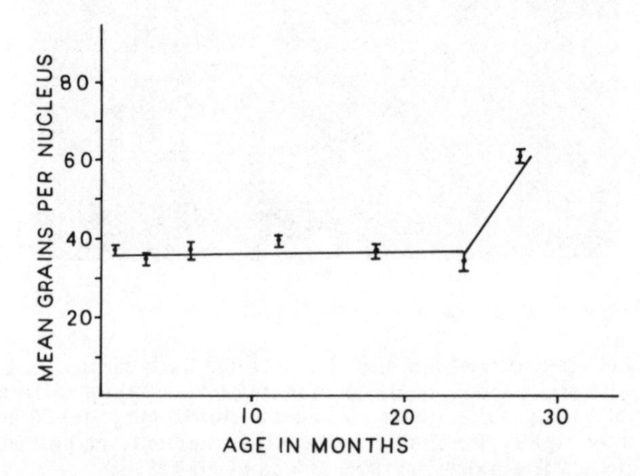

Figure 2. Lifespan related changes in the mean number of grains/nucleus of cortical neurones from layer 3 incubated with terminal transferase and ^{3}H-dGTP. (Data derived from Modak et al, 1985a.)

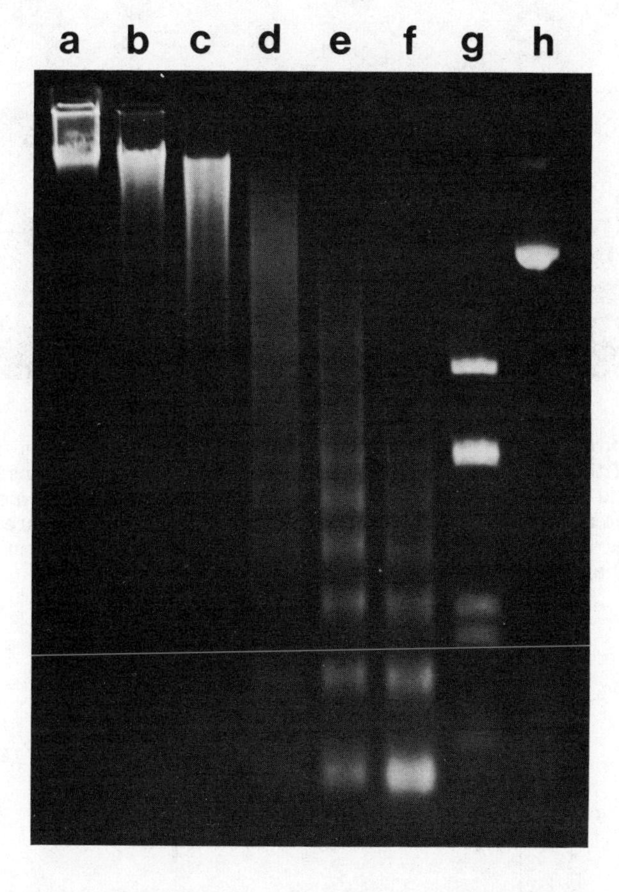

Figure 3: Gel electrophoreogram of 1-month-old liver chromatin-DNA digested with Micrococcal nuclease (4 units/10^6 nuclei) for different times (a) 1 min; (b) 2.5 min; (c) 5 min; (d) 15 min; (e) 30 min; (f) 60 min; (g) SV 40- Hind III restriction fragment; and (h) uncut SV40 DNA. (Data derived from Modak et al, 1985b.)

incorporation remains more or less constant between 1 and 24 months, and then increases dramatically at 28 months.

Structure of chromatin in the ageing mouse liver nuclei

The electrophoretic behaviour of DNA fragments produced by the digestion of mouse liver nuclei with Micrococcal nuclease was examined; an example is shown in Figure 3. From the densitometric tracings, the mean sizes (bp) of different size classes of chromatin-DNA fragments were determined in at least 3 separate experiments at each age-group. The data shown in Table 1 summarize the mean repeat size of DNA in mouse liver chromatin and it can be stated that the chromatin subunit-DNA repeat length does not vary significantly during the period of lifespan examined. We did, however, find that in 28-month-old samples, a small proportion of high MW chromatin DNA resisted attack by the enzyme for about 1h but thereafter these high MW DNA molecules also were degraded.

Table 1. Size of DNA in mouse liver nucleosomes during in vivo ageing*

Age (Months)	Number of Experiments	Estimated DNA size (base pairs) Monomer	Dimer	Trimer	Base pairs/ Repeat unit
1.75	3	207 ± 7	401 ± 14	625 ± 31	207
3.50	3	211 ± 7	408 ± 7	613 ± 14	207
6	5	213 ± 7	407 ± 14	623 ± 12	208
11	4	203 ± 10	407 ± 17	613 ± 16	204
18	3	212 ± 10	410 ± 9	635 ± 4	209
22	1	201	405	608	203
28	5	213 ± 10	412 ± 12	610 ± 22	209
33	1	206	414	620	207
Average		209 ± 8	412 ± 12	619 ± 22	207 ± 3

*Data from Modak et al (1985b)

The time-dependent acid-solubilisation of chromatin-DNA by Micrococcal nuclease reached a plateau at about the 50% level for all samples between 1 and 24 months of age. At 28 and 33 months, the acid precipitable DNA fraction reached a plateau at 58% and 62%, respectively. The age-dependent change in the M. nuclease-resistant DNA fraction is shown in Figure 4 and it seems clear that the chromatin-DNA gains resistance to M. nuclease during the last quarter of the lifespan.

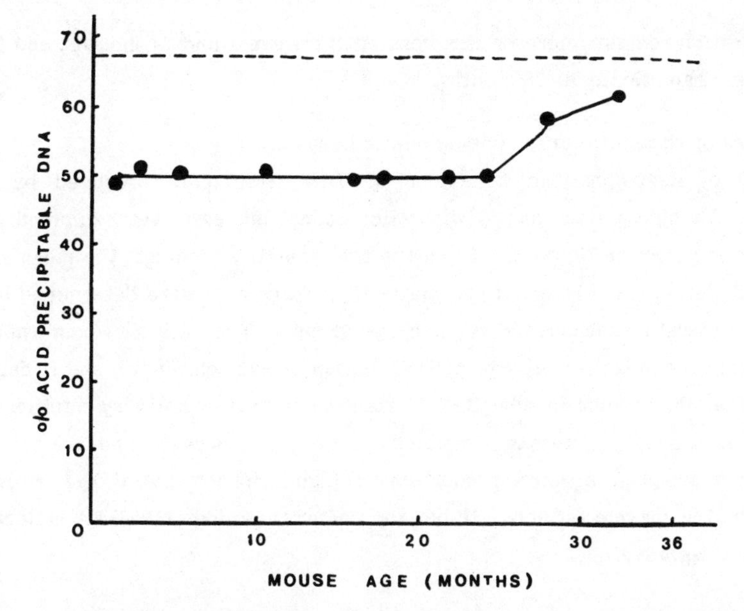

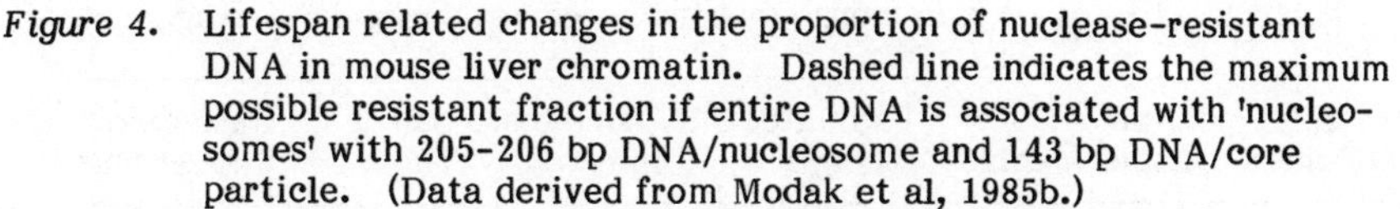
Figure 4. Lifespan related changes in the proportion of nuclease-resistant DNA in mouse liver chromatin. Dashed line indicates the maximum possible resistant fraction if entire DNA is associated with 'nucleosomes' with 205-206 bp DNA/nucleosome and 143 bp DNA/core particle. (Data derived from Modak et al, 1985b.)

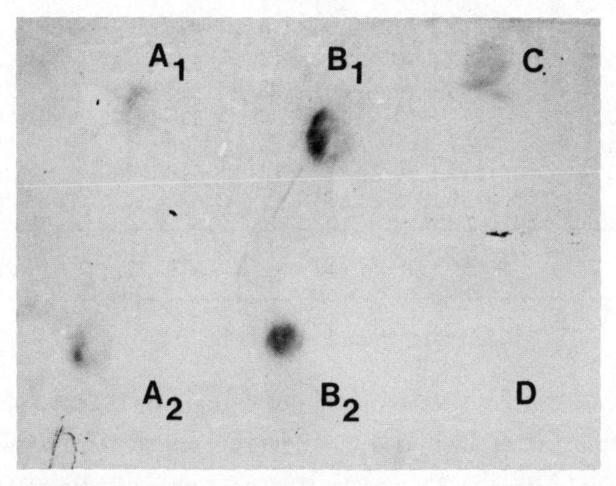

Figure 5. Immunochemical assay for 5-methyl cytosine in DNA. DNA isolated from mouse hepatocyte nuclei was spotted (40 ng in 1 μl) on a BA 85 nitrocellulose paper, dried and treated with antibody to 5 mC and stained by the biotin-avidin-peroxidase technique. A_1 and A_2 contain 25.5-month-old sample. B_1 and B_2 correspond to 11-day-old DNA sample. C is bacteriophage lambda DNA grown in wild type *E.coli*. D is lambda DNA from dcm^- *E.coli* and thus methylation minus.

Immunological detection of 5mC residues in mouse liver DNA

The results of staining various DNA samples by peroxidase-labelled antibodies specific for rabbit anti-mC-antibodies are seen in Figure 5. In all cases 40ng of DNA sample was spotted. As expected, unmethylated lambda DNA showed no colour reaction while the reaction was visible with the methylated lambda DNA sample. Both 11-day-old and 25.5 month-old liver DNA showed positive reactions. Surprisingly, however, the 'young' mouse DNA stained very intensely while the reaction was significantly weaker with 'old' DNA.

DISCUSSION

Denatured DNA can be detected *in situ* by autoradiography of fixed cells incubated with ^{3}H-dNTPs and DNA polymerase alpha (Modak et al, 1969). In terminally differentiating lens fibre cells, nuclei degenerate, lose DNA (Modak & Perdue, 1970) and show greatly increased template activity for DNA polymerase after denaturation of DNA with acid or alkali (Modak et al, 1969; Modak & Bollum, 1970), indicative of an accumulation of strand breaks. The comparison of template activity for this enzyme between 3 month-old 'young' and 30-33 month-old 'old' tissues revealed that 'undenatured' DNA in neurones, hepatocytes, Kupffer cells and cardiac muscle contained 'gapped' DNA in the old cells. Similarly after prior denaturation, the template activity was 4-5 times higher in old cells (Modak & Price, 1971; Price et al, 1971). It was further found that this effect could be mimicked by X-irradiation of 'young' mouse neurones (Modak & Price, 1971). Sedimentation in alkaline sucrose gradients has also revealed an age-dependent decrease in single strand MW DNA in dog cortical neurones (Wheeler & Lett, 1974), chicken erythrocytes and rat muscle cells (Karran & Ormerod, 1973), human diploid cells (Dell'Orco & Whittle, 1981) and human lymphocytes (Turner et al, 1981).

Terminal transferase utilizes free 3'-OH ends in DNA as initiators to catalyse synthesis of homopolymeric deoxynucleotides by end-addition, and it was used to detect such ends *in situ* in chick embryonic lens and adult mouse vagina (Modak, 1972; Modak & Bollum, 1970, 1972; Modak & Traurig, 1972). It was shown that terminal differentiation in lens fibres and keratinizing epithelium is accompanied by a massive increase in free 3'-OH ends. Alkaline sucrose gradient sedimentation of lens fibre DNA showed decreasing single-strand MW (Piatigorsky et al, 1973; Counis et al, 1977) and thus confirmed our findings *in situ*. Recently, the terminal transferase method has been used to demonstrate that gamma rays induce, in a dose-dependent fashion, increasing numbers of free

3'-OH ends in a human colon adenocarcinoma cell line (Fertil et al, 1984). In our studies on the estimation of the relative number of free 3'-OH ends (Modak et al, 1985a), we find that they remain at a constant level between 1 and 24 months of age in mouse cortical neurones (layer 3) and then increase dramatically at 28 months of age. These results confirm our earlier findings using DNA polymerase alpha (Modak & Price, 1971; Price et al, 1971) that the ageing process involves an accumulation of lesions in DNA, and further suggest that the process of strand-break accumulation is not a progressive event but affects genome integrity late during the lifespan in a precipitous manner.

Eukaryotic DNA is completed with histones in the form of chromatin composed of a repeating series of DNA : histone complexes or subunits called nucleosomes (Hewish & Burgoyne, 1973; Olins & Olins, 1974; Noll, 1974). Each nucleosome comprises approximately 200bp-long DNA, of which a stretch of 144bp is wound around and is electrostatically bound to an octamer of histones (2 molecules each of H2A, H2B, H3 and H4) giving rise to the 'core particle'. The points of entry and exit of DNA on the core particle are close to each other and 10bp-long DNA stretches of both are held together by one molecule of histone H_1, the resulting total structure being termed a chromatosome. Adjacent chromatosomes are covalently linked to each other through the DNA fibre which is continuous throughout the length of chromatin; the intervening segment of DNA which is called the 'linker' is approximately 20bp long on either side of the chromatosome. The eukaryote chromatin is thus organized into a linear array of nucleosomes. Linker-DNA is sensitive to both endogenous Ca^{++}/Mg^{++}-dependent endonuclease or exogenous Micrococcal nuclease (Hewish & Burgoyne, 1973; Noll, 1974). DNA repeat-size varies among different tissues of the same organism (Morris, 1976), different cell types from the same organ (Todd & Garrard, 1977), and cells from the same cell lineage (Lohr et al, 1977; Weintraub, 1978). In our studies (Modak et al, 1978; Modak et al, 1985b) it was found that the DNA repeat-length in liver chromatin remained unchanged during the mouse lifespan. Micrococcal (Staphylococcal) nuclease has been reported to digest 50% of chromatin-DNA from adult tissues (Clark & Felsenfeld, 1971; Sollner-Webb & Felsenfeld, 1975; Axel, 1975). The liver nuclei of 1 to 24 month-old mice also contained 50% DNA remaining resistant to the enzyme in a limit-digest, while between 28 and 33 months of age the resistant fraction increased in mice as old as 26-28 months; the results we obtained for 28 month-old livers are not spurious since the trend is maintained even up to 35 months of age (present data, and Modak et al, 1985b). In each nucleosome, 67% DNA is nuclease-

resistant. Even if only 50% were to be resistant, 75% of total DNA can be estimated to be bound to defined nucleosomes. Furthermore, the observation that the resistant fraction increased up to 63% without affecting the size of the DNA repeat suggests that the number of classical nucleosomal structures per unit length DNA may increase late during the lifespan. This latter situation necessarily predicts that the ratio histone:DNA also should increase later during the lifespan. However, determination of these ratios has revealed no such differences (Modak et al, 1985b). Thus it is likely that during the late period pre-existing half-nucleosomal structures (M.nuclease-sensitive) are converted to full nucleosome, which would result in changes in chromatin conformation. A number of reports claiming increased Tm of ageing chromatin (von Hahn, 1970; Pythilla & Sherman, 1968; Zhelabovskaya & Berdyshev, 1972) have been contested by others (Kurtz & Sinex, 1967; Kurtz et al, 1974). From our results it is not possible to state whether the increased resistance to M.nuclease is due to chromatin condensation and/or cross-linking in addition to increased covering of DNA by full nucleosomes; in either case DNA would become decreasingly accessible to interacting nucleases, repair enzymes and DNA/RNA polymerases.

Four major types of DNA repair systems exist (for review, see Hart et al, 1979). In ageing dog neurones, strand-break rejoining activity remains unchanged although these cells accumulate single strand breaks (Wheeler & Lett, 1974). Terminally differentiated lens fibre cells fail to rejoin single strand breaks and to carry out UV-excision repair (Counis et al, 1977; Treton et al, 1981). While in different cell types different types of DNA repair systems may become defective, the loss of genome integrity seems to be due to the decreased accessibility of damaged sites of chromatin-DNA to repair enzymes (Hart et al, 1979; Modak & Unger-Ullmann, 1980). It is known that all damaged sites in DNA are not equally accessible to repair enzymes in eukaryotic chromatin due to the involvement of DNA : histone complexes (Wilkins & Hart, 1973; Bodell, 1977; Cleaver, 1977). Comparison of data on strand breaks in cortical neurones and nuclease-resistance in liver chromatin strongly suggests that the increase in both parameters is temporally coincident. Thus it is hypothesised that the late events during the ageing process involve a decreased accessibility of damaged sites to repair enzymes regardless of the functional status of the latter.

5-methyl cytosine residues in DNA can be quantitatively detected by an immunochemical method (Achwal & Chandra, 1982; Achwal et al, 1984). The preliminary results reported here suggest that there is an appreciably lower number of 5mC residues in 'old' liver DNA. We are now examining samples

derived from tissues at ages distributed throughout the lifespan. Our preliminary data on postmitotic ageing liver cell populations are consistent with those of ageing human diploid fibroblasts (Wilson & Jones, 1983). Increased DNA methylation has generally been correlated with genome inactivity. The decreased methylation reported here may imply that there is an overall nonspecific activation of genes, or that the maintenance methylases are functioning less efficiently, or that the accessibility of DNA methylases decreases. Chromatin template activity for *E coli* RNA polymerases has been shown to decrease in old tissues (Pythilla & Sherman, 1967; Zhelabovskaya & Berdyshev, 1972; Hill, 1976), while no difference could be observed for endogenous RNA polymerase (Hill, 1976). The synthesis of RNA measured by the uptake of ^{3}H-uridine seems to decrease in mouse cerebral cortex, choroid plexus and liver of old animals (Fogg & Pakkenberg, 1981). Thus, the decreased levels of methylation and transcriptional activity seem paradoxical and can be understood by assuming either that these events are only indirectly related, or that the former is due to a decreased accessibility of DNA to methylases. It is also possible that the observed overall decrease in 5mC residues in old liver DNA includes a real methylation of transcribable regions but that it is overshadowed by demethylation of non-transcribed sequences such as satellites and heterochromatic regions.

In prokaryotes, most restriction endonucleases do not cleave at their methylated specific sequences and an undermethylation or demethylation results in DNA degeneration by the endogenous enzymes (Arber, 1974). Nothing is known of eukaryotic restriction endonucleases which, if they were to exist, would result in the introduction of strand breaks thereby decreasing the transcribability of the respective genes. This effect would be further amplified if the accessibility of damaged sites in DNA to repair enzymes decreases. Our results are consistent with this scheme of thinking.

CONCLUSIONS

1. DNA in neurones (layer 3) of the visual area in the cerebral cortex accumulate free 3'-OH ends or single strand breaks during the final quarter of the lifespan in the mouse.
2. Chromatin-DNA from the liver acquires increasing resistance to the Micrococcal nuclease during the final quarter of the lifespan in the mouse.
3. 5-methyl cytosine residues in liver DNA are significantly reduced in old tissue by comparison with young tissue.

4. It is postulated that chromatin changes its structure and conformation in postmitotic tissues during the last quarter of the lifespan of the mouse, thereby decreasing the accessibility of DNA to repair enzymes and methylases.

ACKNOWLEDGMENTS

We gratefully acknowledge the financial support from the Basic Sciences Committee II of the BRNS-Department of Atomic Energy, Science and Engineering Research Council of the Department of Science and Technology, Government of India and the Indian University Grants Commission.

We thank Dr. Dilip N. Deobagkar from the National Chemical Laboratory for critical discussions and Dr. M.V. Joshi for help during the preparation of this manuscript.

REFERENCES

Achwal, C.W. & Chandra, H.S. (1982). A sensitive immunochemical method for detecting 5mC in DNA fragments. FEBS Letters, **150**, 469-472.

Achwal, C.W., Ganguly, P. & Chandra, H.S. (1984). Estimation of amount of 5-methylcytosine in Drosophila melanogaster DNA by amplified ELISA and photoacoustic spectroscopy. EMBO Journal, **3**, 263-266.

Appleby, D.W. & Modak, S.P. (1977). DNA degradation in terminally differentiating lens fibre cells from chick embryos. Proceedings of the National Academy of Sciences, USA, **74**, 5579-5583.

Arber, W. (1974). DNA modification and restriction. In: W. E. Cohn (ed.), Progress in Nucleic Acid Research and Molecular Biology, vol. 14, pp. 1-37. New York: Academic Press.

Axel, R. (1975). Cleavage of DNA in nuclei and chromatin with staphylococcal nuclease. Biochemistry, **14**, 2921-2925.

Bodell, W.J. (1977). Nonuniform distribution of DNA repair in chromatin after treatment with methyl methane sulphonate. Nucleic Acids Research, **4**, 2619-2628.

Clark, R.J. & Felsenfeld, G. (1971). Structure of chromatin. Nature (New Biology Series), **29**, 101-105.

Cleaver, J.E. (1977). Nucleosome structure control rates of excision repair in DNA of human cells. Nature, **270**, 451-453.

Constantini, F.D., Roberts, S., Evans, E.P. & Burtenshaw, M.D. (1984). Position effects and gene expression in the transgenic mouse. In: H.S. Ginsberg & H.J. Vogel (eds.), Transfer and Expression of Eukaryotic Genes, pp. 123-134. New York: Academic Press.

Counis, M.-F., Chaudun, E. & Courtois, Y. (1977). DNA synthesis and repair in terminally differentiating embryonic lens cells. Developmental Biology, **57**, 47-55.

Davidson, E.H. (1976). Gene Activity in Early Development. New York: Academic Press.

Dell'Orco, R.T. & Whittle, W.L. (1981). Evidence for an increased level of DNA damage in high doubling level human diploid cells in culture. Mech. Ageing Dev., **15**, 141-152.

Fertil, B., Modak, S., Chavaudra, N., Debry, H., Meyer, F. & Malaise, E.P. (1984). Detection in situ of -ray induced DNA strand breaks in single cells; Enzymatic labeling of free 3'-OH ends. International Journal of Radiation Research (in press).

Finnegan, D.J., Will, B.A., Bayev, A.A., Bowcock, A.M. & Brown, L. (1981). Transposable DNA sequences in eukaryotes. In: G.A. Dover & R.B. Flavell (eds.), Genome Evolution, pp. 29-40. New York: Academic Press.

Fogg, R. & Pakkenberg, H. (1981). Age-related changes in ^{3}H-uridine uptake in the mouse. Journal of Gerontology, **36**, 680-681.

Gaubatz, J., Ellins, M. & Chalkley, R. (1979). Nuclease digestion studies of mouse chromatin as a function of age. Journal of Gerontology, **34**, 672-679.

Hahn, H.P. von (1970). The regulation of protein synthesis in the aging cells. Experimental Gerontology, **5**, 323-334.

Hart, R.W., D'Ambrosio, S.M., Ng, K.K. & Modak, S.P. (1979). Longevity, stability and repair. Mechanisms of Ageing and Development, **9**, 203-223.

Hewish, D.R. & Burgoyne, L.A. (1973). Chromatin substructure: the digestion of chromatin DNA at regularly spaced sites by a nuclear deoxyribonuclease. Biochemical and Biophysical Research Communications, **52**, 504-510.

Hill, B.T. (1976). Influence of age on chromatin transcription in murine tissues using a heterologous and a homologous RNA polymerase. Gerontology, **22**, 111-123.

Hozumi, N. & Tonegawa, S. (1976). Evidence for somatic rearrangement of Ig genes coding for variable and constant regions. Proceedings of the National Academy of Sciences, USA., **73**, 3628-3632.

Jähner, D., Stuhlmann, H., Stewart, C.L., Harbers, K., Lohler, J., Simon, I. & Jaenisch, R. (1982). De novo methylation and expression of retroviral genomes during mouse embryogenesis. Nature, **298**, 623-628.

Karren, P. & Ormerod, M.G. (1973). Is the ability to repair damage to DNA related to the proliferative capacity of the cell? Biochimica et Biophysica Acta, **299**, 54-64.

Korba, B.E. & Hays, J.B. (1982). Partially deficient methylation of cytosine in DNA at CCA/TGG sites stimulates genetic recombination of bacteriophage lambda. Cell, **28**, 531-541.

Kurtz, D.I. & Sinex, F.N. (1967). Age-related differences in association of brain DNA and nuclear proteins. Biochimica et Biophysica Acta, **145**, 840-842.

Kurtz, D.I., Russell, A.R. & Sinex, F.N. (1974). Multiple peaks in the derivative melting curve of chromatin from animals of varying age. Mechanisms of Ageing and Development, **3**, 37-39.

Lohr, D., Corden, J., Tatchell, K., Kovacic, R.T. & van Holde, K.E. (1977). Comparative subunit structure of HeLa, yeast and chicken erythrocyte

chromatin. Proceedings of the National Academy of Sciences, USA, **74**, 79-83.

McClintock, B. (1956). Controlling elements and the gene. Cold Spring Harbor Symposia on Quantitative Biology, **21**, 197-216.

McMaster, G. & Modak, S.P. (1977). Cellular and biochemical parameters of growth of chick blastoderms during early morphogenesis. Differentiation, **8**, 145-152.

Modak, S.P. (1972). A model for transcriptional control in terminally differentiating lens fiber cells. In: R. Harris, P. Allin & D. Viza (eds.), Cell Differentiation, pp. 339-342. Copenhagen: Munksgaard.

Modak, S.P. & Bollum, F.J. (1970). Terminal lens cell differentiation. III. Initiator activity of DNA during nuclear degeneration. Experimental Cell Research, **62**, 421-432.

Modak, S.P. & Bollum, F.J. (1972). Detection and measurement of single strand breaks in nuclear DNA in fixed lens sections. Experimental Cell Research, **75**, 307-313.

Modak, S.P. & Chalkley, R. (1981). Chromatin structure in Aging. In: R.M. Schimke (ed.), Biological Mechanisms in Aging, pp. 279-289. National Institutes of Health Publication No. 81-2194.

Modak, S.P. & Perdue, S.W. (1970). Terminal lens cell differentiation. I. Histological and microspectrophotometric analysis of nuclear degeneration. Experimental Cell Research, **59**, 43-56.

Modak, S.P. & Price, G.B. (1971). Exogenous DNA polymerase-catalyzed incorporation of deoxynucleotide monophosphates in fixed mouse brain cell nuclei: Changes associated with age and X-irradiation. Experimental Cell Research, **65**, 289-296.

Modak, S.P. & Traurig, H. (1972). Appearance of single strand breaks in the nuclear DNA of terminally differing vaginal keratinizing epithelium. Cell Differentiation, **2**, 351-355.

Modak, S.P. & Unger-Ullmann, C. (1980). Control of genome integrity in terminally differentiating and postmitotic ageing cells, In: R.G. McKinnell, M.A. DiBerardino, M. Blumenfeld & R.D. Bergad (eds.), Differentiation and Neoplasia, pp. 178-190. Berlin/Heidelberg/New York: Springer-Verlag.

Modak, S.P., von Borstel, R.C. & Bollum, F.J. (1969). Terminal lens cell differentiation. II. Template activity of DNA during nuclear degeneration. Experimental Cell Research, **56**, 105-113.

Modak, S.P., Gonet, C., Unger-Ullmann, C. & Chappuis, M. (1978). Chromatin structure in ageing mouse liver. Experientia, **34**, 57.

Modak, S.P., Leuba-Gfeller, G., Failletaz, M.-C. & Modak, M.J. (1985a). Genome integrity in cortical neurons during the life span of the mouse. (Submitted for publication.)

Modak, S.P., Gonet, C., Unger-Ullmann, C. & Chappuis, M. (1985b). Chromatin structure in ageing mouse liver. (Submitted for publication.)

Morris, N.R. (1976). A comparison of the structure of chick erythrocyte and chick liver. Cell, **9**, 627-632.

Noll, M. (1974). Subunit structure of chromatin. Nature, **251**, 249-251.

Olins, A.L. & Olins, D.E. (1974). Spheroid chromatin units (nu bodies). Science, **183**, 330-332.

Piatigorsky, J., Rothschild, S.S. & Milstone, L.M. (1973). Differentiation of lens fibers in explanted embryonic chick lens epithelium. Developmental Biology, **34**, 334-345.

Price, G.B., Modak, S.P. & Makinodan, T. (1971). Age-associated changes in the DNA of mouse tissues. Science, **171**, 917-920.

Pythilla, M.J. & Sherman, F.G. (1968). Age-associated studies on thermal stability and template effectiveness of DNA and nucleoproteins from beef thymus. Biochemical and Biophysical Research Communications, **31**, 340-344.

Radman, M., Wagner, R.E., Glickman, B.W. & Messelson, M. (1980). DNA methylation, mismatch correction and genetic stability. In: M. Alacenic (ed.), Progress in Environmental Mutagenesis, pp. 121-130. New York: Academic Press.

Razin, A. & Riggs, A.D. (1980). DNA methylation and gene function. Science, **210**, 604-609.

Sollner-Webb, B. & Felsenfeld, G. (1975). A comparison of digestion nuclei and chromatin by staphylococcal nuclease. Biochemistry. **14**, 2915-2920.

Spradling, A.C. & Rubin, G.M. (1981). Drosophila genome organization: conserved and dynamic aspects. Annual Review of Genetics, **15**, 219-264.

Todd, R.D. & Garrard, W.T. (1977). Two-dimensional electrophoretic analysis of polynucleosomes. Journal of Biological Chemistry, **252**, 4729-4738.

Treton, J., Modak, S.P. & Courtois, Y. (1981). Analysis of thymidine incorporation in the DNA of chick embryo lens epithelium and lens fibers irradiated with ultraviolet light. Experimental Eye Research, **32**, 61-72.

Turner, D.R., Morley, A.A. & Sheshadri, R.S. (1981). Age-related variations in human lymphocyte DNA. Mechanisms of Ageing and Development, **17**, 305-309.

Weintraub, H. (1978). The nucleosome repeat length increases during erythropoeisis in the chick. Nucleic Acids Research, **5**, 1179-1188.

Wheeler, K.T. & Lett, J.T. (1974). On the possibility that DNA repair is related to age in nondividing cells. Proceedings of the National Academy of Sciences, USA, **71**, 1862-1865.

Wilkins, R.J. & Hart, R.W. (1973). Preferential DNA repair in human cells. Nature (New Biology Series), **247**, 35-36.

Wilson, V.L. & Jones, P.A. (1983). DNA methylation decreases in ageing but not in immortal cells. Science, **220**, 1055-1057.

Zhelabovskaya, S.M. & Berdyshev, G.D. (1972). Composition, template activity and thermostability of liver chromatin in rats of various age. Experimental Gerontology, **7**, 313-332.

INSECTS AS MODELS FOR TESTING THEORIES OF AGEING

M. J. LAMB

Zoology Department, Birkbeck College,
University of London, London, U.K.

human

INTRODUCTION

It seems that until the 1920s the only species for which there was a complete life table was our own (Pearl, 1928). However, in 1921, in the first of a long series of papers with the general title "Experimental Studies on the Duration of Life", Pearl and Parker remedied this situation by providing a complete life table for a second species. This species was an insect, the fruit fly *Drosophila melanogaster*. Raymond Pearl seems to have had few reservations about using an insect as a model for the study of human ageing and longevity; frequently he presented the survival curves for *Drosophila* and man on a single graph to show how similar they are. Subsequently many other workers have studied insect ageing and longevity, and have pointed out the relevance of their studies to ageing in man. Of course, it would be foolish to argue that insects can be an entirely adequate model for the study of human ageing, or even that they can be used for testing all theories of ageing. Quite clearly, they are of little value for testing theories of ageing which, for example, give a primary role to the immune system, or to the neuroendocrine system. However, many theories of ageing should be applicable to all species that show age-dependent deteriorative changes, since they suggest that senescence is the result of time-dependent changes in the cells and molecules which make up all multicellular organisms. For testing these theories, insects should be at least as suitable as any other organism.

What I want to show in this paper is that insects do show ageing changes comparable with those shown by man and other mammals, that they can be used, and have been used, to test theories of ageing, and that some features of some insect species may make them the best organisms to use for testing certain theories.

THE CHOICE OF SPECIES

The first thing to be said about the insects is that they are a very large and diverse group of animals. Potentially they offer not a single test system, but many. Some species have adult lifespans of only a few days (e.g. many Lepidoptera), whereas others, such as some of the Coleoptera, may live for over two years (Rockstein & Miquel, 1973). Within the class there are groups in which the male is the heterogametic sex, and others in which the female is the heterogametic sex; there are species in which feeding takes place throughout adult life, and others such as the silkmoth which have a non-feeding adult stage and can therefore be used to study age-related changes in metabolism in a closed system (e.g. see Osanai, 1983). The hemimetabolous insects have gradual development, whereas holometabolous insects have complete metamorphosis with distinct larval, adult and pupal stages. Since in some adult insects (e.g. *Drosophila*), somatic cell division is almost entirely lacking, it is relatively easy to study the ageing of post-mitotic cells in complete isolation from dividing cells.

In spite of the potential advantages of some of the other insects, most studies of insect ageing have involved relatively few species, mainly the dipteran flies such as *Calliphora*, *Musca* and *Drosophila*, but also some Hymenoptera (e.g. the honey bee and *Habrobracon*), Lepidoptera (e.g. the silkmoth), Dictyoptera (e.g. cockroaches) and Coleoptera (e.g. *Tribolium*, the flour beetle). There are over a million species of insect, but life tables exist for very few of them. The most favoured insect model for ageing studies is still the one that Raymond Pearl used, *D. melanogaster*.

AGEING IN INSECTS

In the laboratory there are a number of ways of demonstrating that adult insects, such as *Drosophila*, show ageing changes. The survival curve (Fig. 1) is of the rectangular type typical of species which show senescence. It starts with a fairly lengthy period during which very few deaths occur, and this is followed by one in which the death rate increases rapidly. The detailed characteristics of the survival curve depend on the strain of flies used and the environmental conditions. For example, temperature, food supply, and population density during both the pre-adult and adult periods all have marked effects on longevity (Lamb, 1978). This dependence of the lifespan on environmental factors has been exploited in attempts to test theories of ageing such as the rate-of-living and free radical theories.

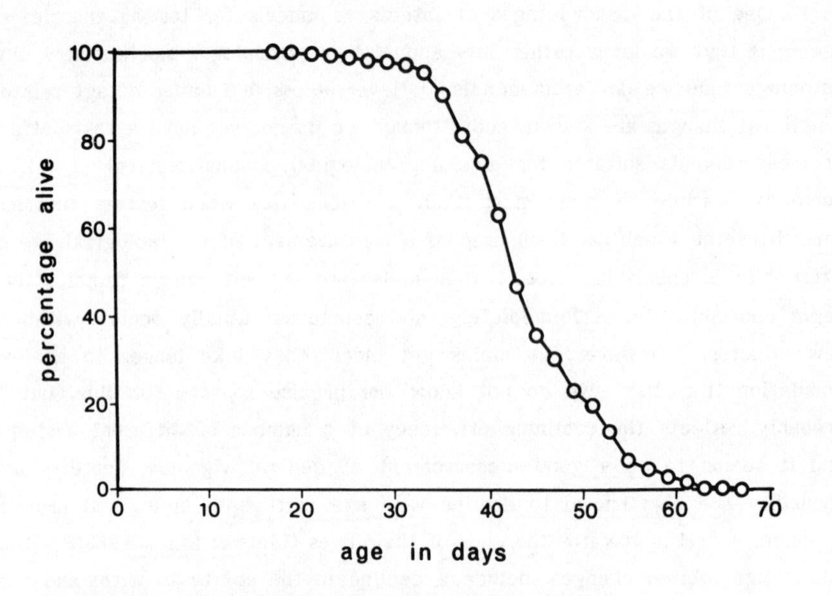

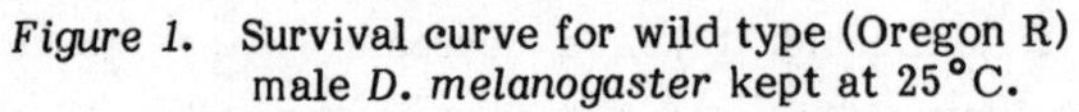

Figure 1. Survival curve for wild type (Oregon R) male *D. melanogaster* kept at 25°C.

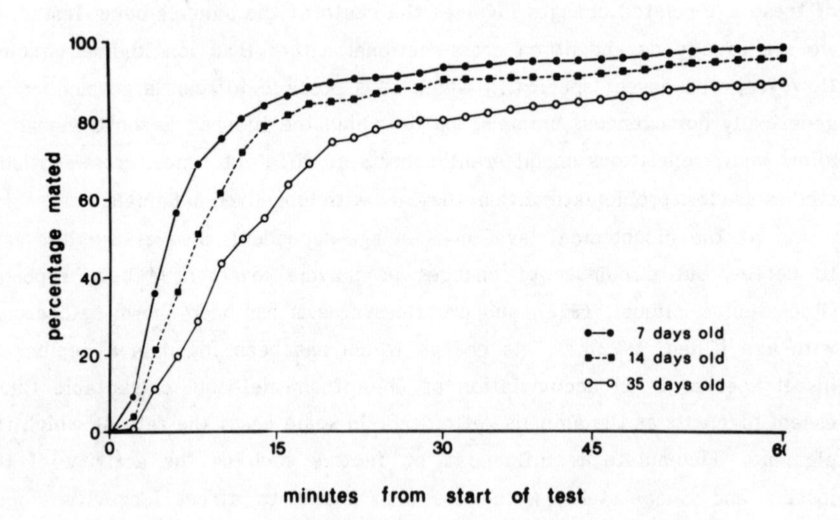

Figure 2. Courtship speed of wild type male *D. melanogaster* of different ages. Individual males were placed with two virgin females at the start of the test.

One of the disadvantages of insects as models for testing theories of ageing is that we know rather less about their physiology, biochemistry, and pathology than we do for mammals. Nevertheless, a number of age-related functional changes are known, and although we do not yet have a test-battery of measurements suitable for assessing biological or physiological age, it is certainly possible to study more than just longevity when testing theories. One character which has been used as a measurement of the biological age of *Drosophila* is courtship speed. When males are put with virgin females they begin courtship almost immediately, and copulation usually occurs within a few minutes. However, as males get older, they take longer to achieve copulation (Fig. 2). We do not know the precise reasons for this, but it probably reflects the declining efficiency of a number of different systems, and it seems to be a good measurement of general vigour. Fertility and fecundity are also known to decline with age, with both males and females becoming infertile towards the ends of their lives (Lints & Lints, 1968). Other known age-related changes include a decline in the ability to withstand high temperatures (Lamb & McDonald, 1973), a decrease in the rate of oxygen consumption (Lints & Lints, 1968), and a decline in the ability to respond to gravity as flies get older (Miquel et al, 1976). Unfortunately, measuring some of these age-related changes involves the death of the animals being tested, so we are usually dependent on cross-sectional rather than longitudinal studies. However, with insect species in which it is possible to rear large numbers of genetically homogeneous animals, and in which the lifespan is short enough to allow many repetitions on different cohorts at different times, cross-sectional studies are less problematical than they are with long-lived mammals.

At the biochemical level marked age-dependent changes are less easy to detect, but a number of changes in enzyme levels have been reported (Rockstein & Miquel, 1973), and protein synthesis has been shown to decrease with age (Chen, 1972). One change which has been found in a number of insect species is an accumulation of chloroform-methanol extractable fluorescent pigments as the animals get older. In some cases the rate at which the pigments accumulate is influenced by factors such as the activity of the insects and temperature, which are also known to affect longevity (Sohal, 1981). The rate of accumulation of these pigments may therefore be a useful indicator of the rate of ageing changes.

The chloroform-methanol extractable fluorescent pigments are believed to be at least in part the same as the so-called age pigments which can be seen

in the cells of many insects and mammals. Recently, Miquel et al (1981) have made a comparative study of age-related structural changes at the cellular and organelle level in *Drosophila* and the mouse. They found that the changes seen in mammalian post-mitotic cells and adult insect cells are very similar indeed. Not only is there an accumulation of age pigment, there is also an increase in size and decrease in number of mitochondria in the cells of post-mitotic tissues in both species, and in both, the number of ribosomes decreases. As a result of their comparative study, Miquel and his colleagues concluded: "In our opinion, *Drosophila* and the mouse share a very fundamental mechanism of aging, namely the age-related disorganization of nonreplicating cells. Mouse senescence is complicated by possible age-related changes in the dividing cells. However, the minimum fine structural alterations found in these replicating cells, even in very old mice, suggest that, as in *Drosophila*, the primary aging changes occur in the fixed postmitotic cells".

THEORIES TESTED

It should be clear from the preceding pages that insects do show distinct age-dependent changes, that a number of these changes can be studied quantitatively, and that some of them are comparable with the changes seen in man and other mammals. Since insects are also generally short-lived and are easy to rear and maintain in large numbers at relatively little cost, many people have found them suitable as models for testing theories of ageing. Table 1 lists some of the theories tested and the insects used. In most cases the theories have also been tested by studying mammals or other organisms, or by using cells in culture, but often insects were chosen because they had special advantages as test organisms. For example, the hymenopteran *Habrobracon* was used for testing the somatic mutation theory because in the Hymenoptera it is possible to obtain individuals of different ploidy. Males are basically haploid and females diploid, but diploid males can be obtained. By showing that untreated haploid and diploid males had essentially the same lifespan, but haploid males were more sensitive to ionising radiation, it was possible for Clark et al (1963) to conclude that classical mutations occurring in somatic cells are unlikely to be the cause of ageing. In many of the other studies listed in the Table, *Drosophila* has been the test animal. Part of the reason for the popularity of *Drosophila* is no doubt its availability to research workers, since it is kept in biology departments throughout the world, but it also has another advantage. Raymond Pearl was well aware of this. He said: "This organism has the great advantage over any other

Table 1. Some theories tested using insect models

Theories	Insects used	Reference
Evolutionary	*Tribolium*	Sokal, 1970; Mertz, 1975
	Drosophila	Rose & Charlesworth, 1980
Rate-of-living	*Drosophila*	Pearl, 1928; Trout & Kaplan, 1970; Miquel et al, 1976
	Musca	Sohal, 1981
	Oncopeltus	McArthur & Sohal, 1982
	Aedes	Nayar, 1972
Developmental	*Drosophila*	Lints & Soliman, 1977
	Tribolium	Lints & Soliman, 1977
Somatic mutation	*Habrobracon*	Clark et al, 1963
	Drosophila	Lamb, 1965
Error catastrophe	*Drosophila*	Harrison & Holliday, 1967; Bozcuk, 1976; Parker et al, 1981
Free radical	*Drosophila*	Sheldahl & Tappel, 1974; Miquel et al, 1983
	Musca	Sohal, 1981
	Oncopeltus	McArthur & Sohal, 1982

which could be used, that its genetic behaviour and potentialities are more thoroughly understood than those of any other animal". This was written in 1921, but it is still true. In my opinion, it is the main justification for using this rather small insect for testing theories of ageing, and why in some cases it may be the model of choice. It is possible to produce flies which have built-in defects and see what effect this has on the rate of ageing. One example of this is the use of the high activity 'shaker' mutants by Trout and Kaplan (1970) to test the rate-of-living theory. They were able to show that high activity was correlated with a shorter lifespan and higher metabolic rate, and thus provided data supporting the theory. There are many other opportunities to make this kind of hypothesis-testing using *Drosophila*, particularly now that so many biochemical mutants are becoming available. For example, hypotheses which suggest that DNA repair is important in determining the rate of ageing should be

testable with *Drosophila* as more becomes known about repair-deficient mutants in this species. Some work using these mutants has already been published (Miquel et al, 1983), and more should follow.

A GERM LINE REJUVENATION HYPOTHESIS

Recently, in my laboratory, we have been using *Drosophila* as a model for testing a hypothesis about ageing which suggests that the amount of DNA repair in germ line cells is important. In our experiments we are making use of some rather unusual strains of flies, and are also exploiting one peculiarity of *Drosophila* which makes it particularly suitable for testing the hypothesis. The hypothesis is concerned with the immortality of the germ line. As Weismann realised over a hundred years ago, whereas most organisms age and die, the germ cell line is passed on from generation to generation and thus is in some senses 'immortal'. What is it that makes the germ line potentially immortal? In particular, what special features maintain the integrity of the genetic material? A number of suggestions have been made (for a review see Medvedev, 1981), but the idea that interests us is that rejuvenation takes place during meiotic recombination. Traditionally, sexual reproduction and genetic recombination have been thought of mainly as mechanisms which generate diversity and allow favourable mutations occurring in different individuals to be brought together. However, evolutionary biologists have always had difficulty in understanding the nature of the selective forces that are involved in the origin and maintenance of sexual reproduction and recombination (Maynard Smith, 1978). Recently several people independently have suggested that sex and recombination may have evolved in the first place in response to selection for processes which repair DNA damage, and that DNA repair is still one of the most important functions of meiotic recombination (Bernstein, 1977; Martin, 1977; Walker, 1978). These authors point out that certain types of damage which involve both of the complementary strands of the DNA molecule cannot be repaired unless an undamaged template is available for copying. During meiosis, homologous chromosomes pair very closely, and this provides an opportunity for the damaged molecule to use the DNA of the homologous chromosome as a template for its own repair. It is argued that the recombinational repair which takes place in meiotic prophase rejuvenates the germ line; somatic cells age and die because they are unable to repair some types of DNA damage.

There is little direct evidence for the hypothesis that recombinational repair is rejuvenatory. One of the most interesting pieces of indirect evidence

comes from work with the ciliate *Paramecium*. This protozoan can reproduce both sexually and asexually. If it is prevented from undergoing sexual reproduction, it shows clonal senescence, i.e. after several hundred mitotic divisions the clone eventually dies out. If, on the other hand, the strains are allowed to go through a sexual process, either autogamy or conjugation, they are restored (Siegel, 1967). Thus, the sexual process in this organism is rejuvenatory.

The reason why we thought that *D. melanogaster* might be suitable for testing the hypothesis of germ line rejuvenation through recombinational repair is that the males of this species have the rather unusual feature of showing no genetic recombination. Although during meiosis the chromosomes pair, the association is not the same as in most other organisms. Cytologically, no chiasmata are seen during meiosis, and at the ultrastructural level, synaptonemal complexes, which are believed to be important in mediating genetic exchange, are not present (Baker & Hall, 1976). On the assumption that the lack of recombination and absence of chiasmata and synaptonemal complexes in *Drosophila* males mean that recombinational repair does not take place, we decided to see what would happen to longevity if part of the genome of flies was passed continually through males. If recombinational repair is necessary to maintain the DNA damage-free, and if damaged DNA leads to a decrease in the efficiency of animals containing it, then flies in which some of the chromosomes have always been passed through the males might be expected to have shorter lifespans.

In order to pass chromosomes continually through males, we made use of strains constructed by Novitski (1976) which contain what are known as entire compound second chromosomes (C(2)EN). In these strains, the second chromosomes, autosomes which constitute more than a third of the genome of *D. melanogaster*, are both attached to a single centromere (Fig. 3). Consequently, instead of carrying a single copy of chromosome-II, the gametes produced from C(2)EN flies carry either two copies of the chromosome or none. When crossed with normal flies, no viable progeny are produced since all the zygotes are aneuploid. However, when females with C(2)EN chromosomes are crossed with males which also carry the compound chromosome, half of the zygotes are expected to survive. The inviable ones are those containing four copies or no copies of chromosome-II. The viable ones result from fusion of a gamete containing the compound and one without; half of the viable zygotes should carry the maternal compound, and the other half the paternal compound. Thus, in theory, 50% of the progeny will inherit both second

chromosomes from their mother (matroclinous offspring), and 50% will inherit both chromosomes from their father (patroclinous offspring). By using genetic markers on the second chromosomes, it was possible to maintain lines in which both copies of chromosome-II were passed continually through males for a number of generations. If the assumption about the lack of recombinational repair in male *D. melanogaster* is correct, and if recombinational repair is essential to rejuvenate the germ line, the progeny which contain these male-transmitted second chromosomes would be expected to carry large amounts of unrepaired damage and therefore have shorter lifespans.

Table 2. Average lifespans of males in which both second chromosomes were inherited from either the male parent or female parent. In the WM line the males transmitted chromosomes carrying wild type alleles, in the MM line males transmitted chromosomes carrying the mutant markers *brown* eyes and *black* body colour.

	Wild type males			Mutant males		
Line and generation	no.	mean lifespan (days)	% dead by day 25	no.	mean lifespan (days)	% dead by day 25
0	100	51.7 ± 1.59	8.0	150	50.0 ± 1.44	12.0
WM 1	60	57.4 ± 2.34	8.3	110	56.8 ± 1.52	5.5
2	68	60.9 ± 2.03	4.4	100	58.3 ± 1.28	2.0
3	70	59.9 ± 1.88	5.7	100	57.7 ± 1.73	7.0
4	70	57.5 ± 1.78	5.7	100	49.3 ± 1.21	4.0
5	99	55.7 ± 1.65	7.1	97	49.7 ± 1.56	8.2
6	99	45.9 ± 1.19	2.0	100	45.8 ± 1.40	6.0
8	99	47.2 ± 1.54	9.1	100	48.5 ± 1.31	4.0
9	100	43.2 ± 1.46	10.0	100	50.8 ± 1.56	8.0
10	100	43.5 ± 1.24	10.0	100	44.1 ± 1.25	6.0
11	100	40.5 ± 1.48	14.0	98	49.6 ± 1.48	10.2
12	100	41.0 ± 1.62	15.0	100	45.9 ± 1.62	11.0
14	100	51.3 ± 1.86	12.0	100	54.3 ± 1.40	3.0
15	90	50.6 ± 1.88	7.8	90	54.7 ± 1.69	10.0
MM 1	110	51.5 ± 1.39	5.5	100	58.0 ± 1.80	7.0
2	90	57.2 ± 1.42	1.1	90	57.0 ± 1.63	5.6
3	100	55.0 ± 1.50	3.0	90	56.2 ± 1.61	4.4
4	100	50.1 ± 1.13	5.0	50	45.3 ± 2.28	12.0
5	99	47.8 ± 1.07	5.1	100	48.4 ± 1.49	10.0
6	100	41.0 ± 1.15	8.0	90	40.4 ± 1.34	8.9
8	100	42.0 ± 1.29	10.0	100	42.2 ± 1.35	10.0
9	100	34.8 ± 1.05	12.0	100	40.6 ± 1.63	18.0
10	100	38.5 ± 1.34	19.0	97	41.8 ± 1.67	20.6
11	100	40.4 ± 1.26	13.0	100	42.5 ± 1.52	14.0
12	100	42.9 ± 1.50	12.0	100	40.3 ± 1.68	23.0
14	100	48.9 ± 1.42	6.0	100	56.4 ± 1.63	5.0
15	100	50.5 ± 1.13	2.0	100	50.9 ± 1.74	8.0

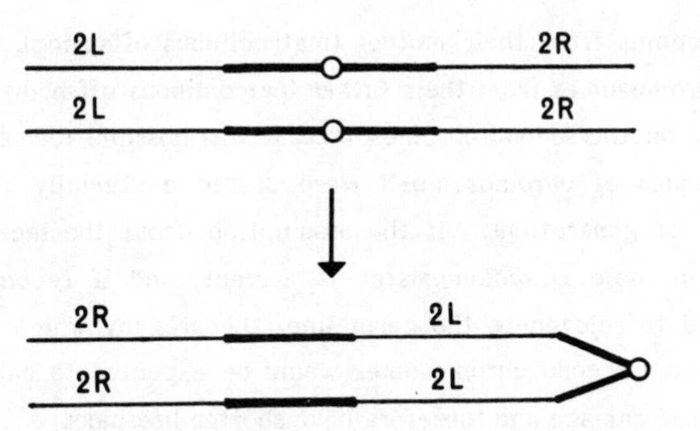

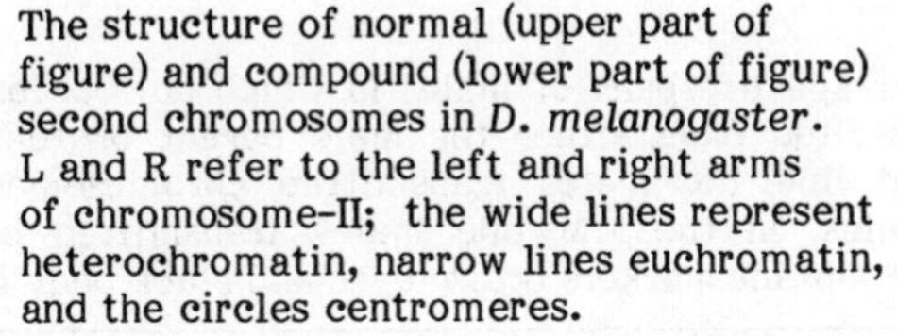

Figure 3. The structure of normal (upper part of figure) and compound (lower part of figure) second chromosomes in *D. melanogaster*. L and R refer to the left and right arms of chromosome-II; the wide lines represent heterochromatin, narrow lines euchromatin, and the circles centromeres.

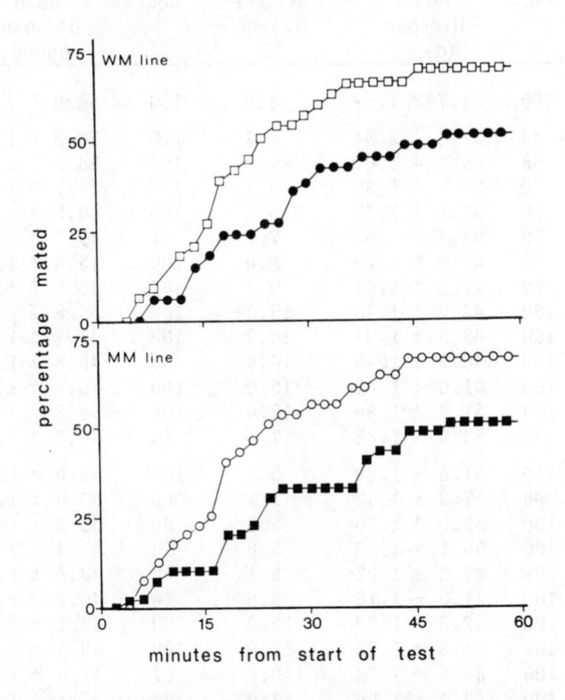

Figure 4. Courtship success of male *D. melanogaster* after 19 generations in which both second chromosomes have been transmitted through the male parent (solid symbols) or female parent (open symbols). Circles are wild type flies, squares are mutant.

Two lines have been maintained. In one, the wild type male line (WM), the males used to continue the line have C(2)EN chromosomes carrying wild type alleles, whereas the females are homozygous for the markers *black* body colour and *brown* eyes. In the other line, the mutant male line (MM), the situation is reversed: the males are homozygous for the markers *black* and *brown*, and the females are wild type. Both lines have now been maintained for over 20 generations. From each line in each generation we are taking matroclinous and patroclinous males and are comparing their lifespans or ageing characteristics.

Table 2 gives the mean lifespans of the two types of male in each line. So far, the results provide no clear evidence for patroclinous males having shorter the patroclinous males die young. Interestingly, the patroclinous males also seem to be less vigorous. The data are not yet very extensive, but we have found in the three generations where it has been studied that the courtship success of young patroclinous males is lower than that of their matroclinous siblings (Fig. 4).

Although the data from the longevity studies do not provide much support for the theory that recombinational repair is necessary to rejuvenate the germ line, other results from the experiment suggest that the theory may be correct. Figure 5 shows the percentage of patroclinous progeny in each generation.

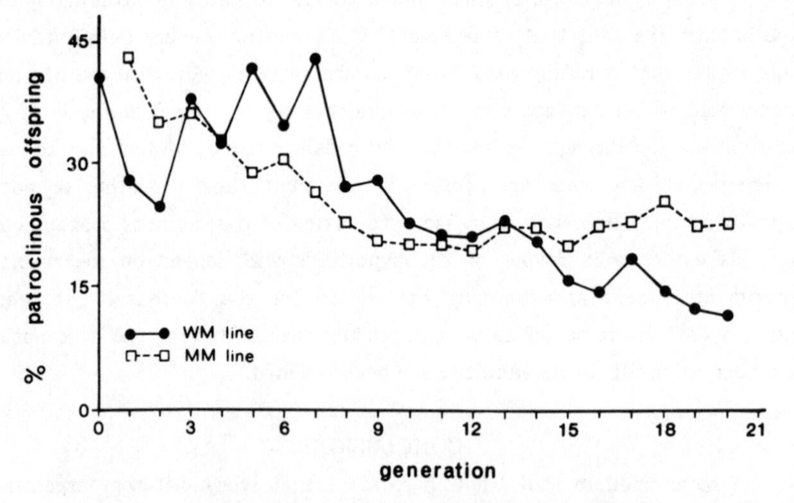

Figure 5. Percentage of the adult offspring in which the second chromosomes came from the male parents. ●—● male parents wild type (WM line); □--□ male parents mutant (MM line).

Instead of the expected 50%, fewer were obtained, and the proportion has tended to decrease over subsequent generations. This decrease in the proportion of progeny receiving the compound chromosome from the male parent has been reported previously, and given various interpretations (Novitski et al, 1981; Strommen, 1982). However, it is also possible to interpret the data in terms of the theory being tested here. It may be that fewer and fewer patroclinous progeny are produced because those chromosomes which carry a heavy load of unrepaired damage are being eliminated before they form part of the genome of an adult fly. Medvedev (1981) and Reanney et al (1983) have stressed that there is a very effective screen or filter operating during gametogenesis and embryogenesis which prevents defective genetic information being passed to the next generation. During the haploid stage, only the most viable genomes are likely to survive; during embryonic and larval development, progeny with inadequate genetic information will die. In C(2)EN stocks, Novitski et al (1981) found that sperm maturation is not normal; in my laboratory we have found that the proportion of patroclinous progeny decreases with the length of time the sperm has been stored in the reproductive tract of the female. It seems likely, therefore, that some chromosomes transmitted from the father are selectively eliminated in the gamete stage.

Whatever the cause of the reduced number of patroclinous progeny in our experiments, the fact that some selective elimination occurs before the adult stage means that our failure to detect a substantial reduction in the lifespan of patroclinous males cannot be taken as evidence against the hypothesis of germ line rejuvenation through recombinational repair. It may be that only relatively undamaged chromosomes are present in our adult flies. Whether or not the observation that there is a decreased proportion of patroclinous progeny can be taken as evidence in favour of the hypothesis will depend on the results of experiments aimed at examining the reason for the decrease. At present, although the system seems to be a promising one for testing the hypothesis, no firm conclusions about its validity have been reached.

CONCLUSIONS

I have tried to show that in spite of their very different organisation, insects do show ageing changes comparable with those seen in mammals, and are adequate models for testing many theories of ageing. Some insect species possess peculiarities which may make them ideal for studying some aspects of ageing. In particular, because of the availability of mutant strains and the

wealth of knowledge about its genetics, the insect *D. melanogaster* may be the best model to choose for testing some theories.

REFERENCES

Baker, B.S. & Hall, J.C. (1976). Meiotic mutants: genic control of meiotic recombination and chromosome segregation. In: M. Ashburner & E. Novitski (eds.), The Genetics and Biology of Drosophila, vol. 1a, pp. 352-434. London: Academic Press.

Bernstein, H. (1977). Germ line recombination may be primarily a manifestation of DNA repair processes. Journal of Theoretical Biology, **69**, 371-380.

Bozcuk, A.N. (1976). Testing the protein error hypothesis of ageing in Drosophila. Experimental Gerontology, **11**, 103-112.

Chen, P.S. (1972). Amino acid pattern and rate of protein synthesis in aging Drosophila. In: M. Rockstein & G.T. Baker (eds.), Molecular Genetic Mechanisms in Development and Aging, pp. 199-226. New York/London: Academic Press.

Clark, A.M., Bertrand, H.A. & Smith, R.E. (1963). Life span differences between haploid and diploid males of Habrobracon serinopae after exposure as adults to X-rays. American Naturalist, **97**, 203-208.

Harrison, B.J. & Holliday, R.L (1967). Senescence and the fidelity of protein synthesis in Drosophila. Nature, **213**, 990-992.

Lamb, M.J. (1965). The effects of X-irradiation on the longevity of triploid and diploid female Drosophila melanogaster. Experimental Gerontology, **1**, 181-187.

Lamb, M.J. (1978). Ageing. In: M. Ashburner & T.R.F. Wright (eds.), The Genetics and Biology of Drosophila, vol. 2c, pp. 43-104. London: Academic Press.

Lamb, M.J. & McDonald, R.P. (1973). Heat tolerance changes with age in normal and irradiated Drosophila melanogaster. Experimental Gerontology, **8**, 207-217.

Lints, F.A. & Lints, C.V. (1968). Respiration in Drosophila, II: Respiration in relation to age by wild, inbred and hybrid Drosophila melanogaster imagos. Experimental Gerontology, **3**, 341-349.

Lints, F.A. & Soliman, M.H. (1977). Growth rate and longevity in Drosophila melanogaster and Tribolium castaneum. Nature, **266**, 624-625.

Maynard Smith, J. (1978). The Evolution of Sex. London: Cambridge University Press.

Martin, R. (1977). A possible genetic mechanism of aging, rejuvenation, and recombination in germinal cells. In: R.S. Sparkes, D.E. Comings & C.F. Fox (eds.), ICN-UCLA Symposia on Molecular and Cellular Biology, vol. 7: Molecular Human Cytogenetics, pp. 355-373. New York: Academic Press.

McArthur, M.C. & Sohal, R.S. (1982). Relationship between metabolic rate, aging, lipid peroxidation, and fluorescent age pigment in milkweed bug, Oncopeltus fasciatus. Journal of Gerontology, **37**, 268-274.

Medvedev, Z.A. (1981). On the immortality of the germ line: genetic and biochemical mechanisms. A review. Mechanisms of Ageing and Development, **17**, 331-359.

Mertz, D.B. (1975). Senescent decline in flour bettle strains selected for early adult fitness. Physiological Zoology, **48**, 1-23.

Miquel, J., Lundgren, P.R., Bensch, K.G. & Atlan, H. (1976). Effects of temperature on the life span, vitality and fine structure of Drosophila melanogaster. Mechanisms of Ageing and Development, **5**, 347-370.

Miquel, J., Economos, A.C. & Bensch, K.G. (1981). Insect vs. mammalian aging. In: J.E. Johnson (ed.), Aging and Cell Structure, vol. 1, pp. 347-379. New York: Plenum.

Miquel, J., Binnard, R. & Fleming, J.E. (1983). Role of metabolic rate and DNA-repair in Drosophila aging: implications for the mitochondrial mutation theory of aging. Experimental Gerontology, **18**, 167-171.

Nayar, J.K. (1972). Effects of constant and fluctuating temperatures on life span of Aedes taeniorhynchus adults. Journal of Insect Physiology, **18**, 1303-1313.

Novitski, E. (1976). The construction of an entire compound two chromosome. In: M. Ashburner & E. Novitski (eds.), The Genetics and Biology of Drosophila, vol. 1b, pp. 562-568. London: Academic Press.

Novitski, E., Grace, D. & Strommen, C. (1981). The entire compound autosomes of Drosophila melanogaster. Genetics, **98**, 257-273.

Osanai, M. (1983). Lifespan and protein synthesis in several Bombyx mori mutants with genetic abnormalities in amino acid and protein metabolism. Experimental Gerontology, **18**, 383-391.

Parker, J., Flanagan, J., Murphy, J. & Gallant, J. (1981). On the accuracy of protein synthesis in Drosophila melanogaster. Mechanisms of Ageing and Development, **16**, 127-139.

Pearl, R. (1928). The Rate of Living. London: London University Press.

Pearl, R. & Parker, S.L. (1921). Experimental studies on the duration of life. I. Introductory discussion of the duration of life in Drosophila. American Naturalist, **55**, 481-509.

Reanney, D.C., MacPhee, D.G. & Pressing, J. (1983). Intrinsic noise and the design of the genetic machinery. Australian Journal of Biological Sciences, **36**, 77-91.

Rockstein, M. & Miquel, J. (1973). Aging in Insects. In: M. Rockstein (ed.), The Physiology of Insecta, 2nd edn., vol. 1, pp. 371-478. New York/London: Academic Press.

Rose, M. & Charlesworth, B. (1980). A test of evolutionary theories of senescence. Nature, **287**, 141-142.

Sheldahl, J.A. & Tappel, A.L. (1974). Fluorescent products from aging Drosophila melanogaster: an indicator of free radical lipid peroxidation damage. Experimental Gerontology, **9**, 33-41.

Siegel, R.W. (1967). Genetics of ageing and the life cycle in ciliates. Symposia of the Society for Experimental Biology, **21**, 127-148.

Sohal, R.S. (1981). Metabolic rate, aging, and lipofuscin accumulation. In: R.S. Sohal (ed.), Age Pigments, pp. 303-316. Amsterdam: Elsevier/North

Holland Biomedical Press.

Sokal, R.R. (1970). Senescence and genetic load: evidence from Tribolium. Science, **167**, 1733-1734.

Strommen, C.A. (1982). Paternal transmission of entire compounds of chromosome two in Drosophila melanogaster. Molecular and General Genetics, **187**, 126-131.

Trout, W.E. & Kaplan, W.D. (1970). A relation between longevity, metabolic rate, and activity in shaker mutants of Drosophila melanogaster. Experimental Gerontology, **5**, 83-92.

Walker, I. (1978). The evolution of sexual reproduction as a repair mechanism. Part 1. A model for self-repair and its biological implications. Acta Biotheoretica, **27**, 133-158.

HUMAN CELL CULTURE SYSTEMS IN THE STUDY OF AGEING

A. H. BITTLES[1] and Y. SAMBUY[2]

[1] Department of Anatomy and Human Biology, King's College, University of London, London, U.K.
[2] Istituto di Anatomia ed Istologia Patologica II, Universita Degli Studi di Roma, Rome, Italy

INTRODUCTION

The first successful tissue culture was reported 100 years ago when the medullary plate from a chick embryo was successfully maintained in warm saline for 4 to 5 days (Roux, 1885). However, it was not until the introduction of methods based initially on lymph (Harrison, 1907) and later plasma clots (Carrel & Burrows, 1910) that cell culture, as opposed to cell maintenance, became established as a reproducible technique. The link between *in vitro* culture and lifespan studies was forged at an early stage following the claim that tissue grown in the laboratory was potentially immortal (Carrel, 1912, 1914; Ebeling, 1922). This belief persisted until it was demonstrated that human fibroblast cultures had a finite, reproducible lifespan (Hayflick & Moorhead, 1961; Hayflick, 1965), a finding which primarily was responsible for the rapid adoption of cell culture methods in the study of ageing. Evidence that the lifespan of diploid fibroblasts *in vitro* was governed by the number of their cell divisions rather than absolute time in culture (Dell'Orco et al, 1973; Goldstein & Singal, 1974; Harley & Goldstein, 1978) served to further emphasize the potential of cell cultures in this field of research.

A number of authors have disputed the validity of fibroblast models in the study of ageing on theoretical and practical grounds. For example, it has been proposed that the loss of division potential exhibited by ageing fibroblasts represents differentiation into a non-cycling stage rather than senescence *per se* (Bell et al, 1978). As terminal differentiation is a pre-requisite for cellular ageing (Walton, 1982), this view would appear to be more a matter of semantics than a point of major import (Hayflick, 1980a), but some authors still appear to regard it as important (Reiner, 1983; Miquel & Fleming, 1984). Criticism aimed specifically at fibroblast cultures because of the cells' reverting, post-mitotic role *in vivo* (Kohn, 1982) appears equally mis-directed. Many other human cell types have been grown in the laboratory, for example, arterial smooth muscle

cells (Bierman, 1978), epidermal keratinocytes (Rheinwald & Green, 1975; Gilchrest, 1983), glial cells (Ponten et al, 1983; Thaw et al, 1984) and liver cells (Le Guilly et al, 1973; Kahn et al, 1977a). All exhibit a finite lifespan *in vitro*, with age-related morphological changes similar to those in fibroblasts. Finally, the suggestion that the finite *in vitro* lifespan of human diploid cells may be an experimental artefact associated with the somewhat limiting scale of the culture conditions (Kirkwood & Holliday, 1975; Holliday et al, 1977) seems improbable. Despite the large number of investigators who have cultured specific, well-characterised human cell strains, such as WI38 and MRC5 during the course of the last twenty years, little evidence in support of the theory has been forthcoming (Hayflick, 1980a).

GROWTH CHARACTERISTICS AND MORPHOLOGY OF SENESCENT, CULTURED FIBROBLASTS

Diploid fibroblasts, derived either from adult skin or fetal skin and/or lung, have been the human cells most commonly employed in ageing studies. The main growth characteristics cited in support of their use are listed in Table 1. Once established in culture, the cells display age-related morphological changes and, in terms of their applicability as a model of ageing, it is important to note that the changes in late passage cells are similar in many respects to those present in fibroblasts ageing *in vivo* (Hayflick, 1980b; Walton, 1982; Johnson, 1984; Pierragi et al, 1984). Although older fibroblasts tend to be more

Table 1. Growth characteristics of senescent cultured fibroblasts

	References
Inverse relationship between age of the explant donor and the number of cell doublings *in vitro*.	Martin et al (1979) Schneider & Mitsui (1976)
Lengthened latent period associated with cell outgrowth from explants in tissue from old donors.	Waters & Walford (1970)
Decrease in colony size obtained from the explants of old donors.	Smith et al (1977)
Reduced plating efficiency of cells from older donors.	Vracko et al (1983)
Reduced in vitro lifespan of explants taken from individuals with inherited disorders.	Goldstein (1969) Holliday et al (1974) Goldstein et al (1979) Shapiro et al (1979)

heterogeneous with respect to size and shape (Lipetz & Cristofalo, 1972; Wolosewick & Porter, 1977), on average there is decreased saturation density in older cultures (Macieira-Coelho et al, 1966a,b) accompanied by an increase in cell size (Simons & van der Broeck, 1970; Bowman et al, 1975; Greenberg et al, 1977) and cell volume (Mitsui & Schneider, 1976b; Schneider & Fowlkes, 1976).

The application of scanning and transmission electron microscopy to the study of ageing has revealed an extensive range of surface and membrane features that change with the onset of senescence in cultured human diploid fibroblasts (Kelley & Vogel, 1984). There is a decrease in the overall charge of the cells (Bosmann et al, 1975) and a reduction in the surface-associated fibronectin network thought to be implicated in cell adhesion and spreading (Aizawa et al, 1980; Vogel et al, 1981). The distribution of intra-membrane particles partially reverses with age (Kelley & Skipper, 1977) and a decrease has been observed in the number and organisation of gap junctions involved in the exchange of ions and metabolites between cells (Kelley, 1976). Even components of the cytoskeleton show evidence of altered organisation with increasing lifespan in culture (Kelley et al, 1980).

Transmission electron microscopy has been used to determine age-related changes in the ultra-structure of cellular organelles, summarised in Table 2. A number of the typical features associated with fibroblast ageing *in vitro* are seen in Figure 1, an electron micrograph of a human embryonic lung fibroblast taken two cell population doublings (CPD) before the cell strain ceased to divide in culture. Although scanning and transmission electron microscopy studies can provide much useful information on the characteristics of senescent cells, they contribute relatively little to our understanding of the underlying causes. In addition, the apparent changes with increased sub-cultivation may represent sub-optimal, experimental conditions rather than genuine, age-related effects (Pool & Metter, 1984).

THEORIES OF CELLULAR AGEING

Many theories have been advanced to explain ageing (reviewed by Shock, 1981) but most current investigators have concentrated on two alternative interpretations of the experimental findings obtained with cultured cells; the infidelity of information flow due to faulty transcription and/or translation (Orgel, 1963, 1970, 1973) or post-translational modification (Gershon & Gershon, 1976; Rothstein, 1977). Theoretical examples of how each mechanism might function have been extensively considered (Gershon, 1979; Hirsch, 1983).

Table 2. Age-related organelle changes in human diploid fibroblasts

Change	Reference
Nucleus	
Larger nucleus with increased invagination	Mitsui & Schneider (1976a) Wolosewick & Porter (1977) Lee et al (1978)
Increased condensation of chromatin on nuclear membrane	Johnson (1979)
No change in chromatin condensation	Lipetz & Cristofalo (1978) Basler et al (1979) Boak et al (1982)
Decreased total number of nucleoli; Increased proportion of cells with one large nucleolus; Increased mean nucleolar area and dry mass	Bemiller & Lee (1976) Bemiller & Miller (1979)
Cytoplasm	
Endoplasmic reticulum	
Decrease in total rough endoplasmic reticulum (RER)	Johnson (1979)
Constricted RER	Lipetz & Cristofalo (1972) Johnson (1979) Mitsui et al (1980)
Dilated RER	Basler et al (1979)
Mitochondria	
Fewer mitochondria with completely transverse cristae	Lipetz & Cristofalo (1972)
Reduction in total number and increased proportion with bizarre morphology	Johnson (1979)
Golgi apparatus	
More developed	Lipetz & Cristofalo (1972)
Less developed	Boak et al (1982)
No change	Robert (1977)
Lysosomes and vacuolar bodies	
Increase in number and size	Robbins et al (1970) Lipetz & Cristofalo (1972) Johnson (1979)

Experimental studies generally have centred either on the specific activities of enzyme proteins at different stages of their *in vitro* lifespan or on properties which may reveal changes in the structure and/or organisation of cellular macromolecules (Dreyfus et al, 1983; Gracy et al, 1983), including the relative

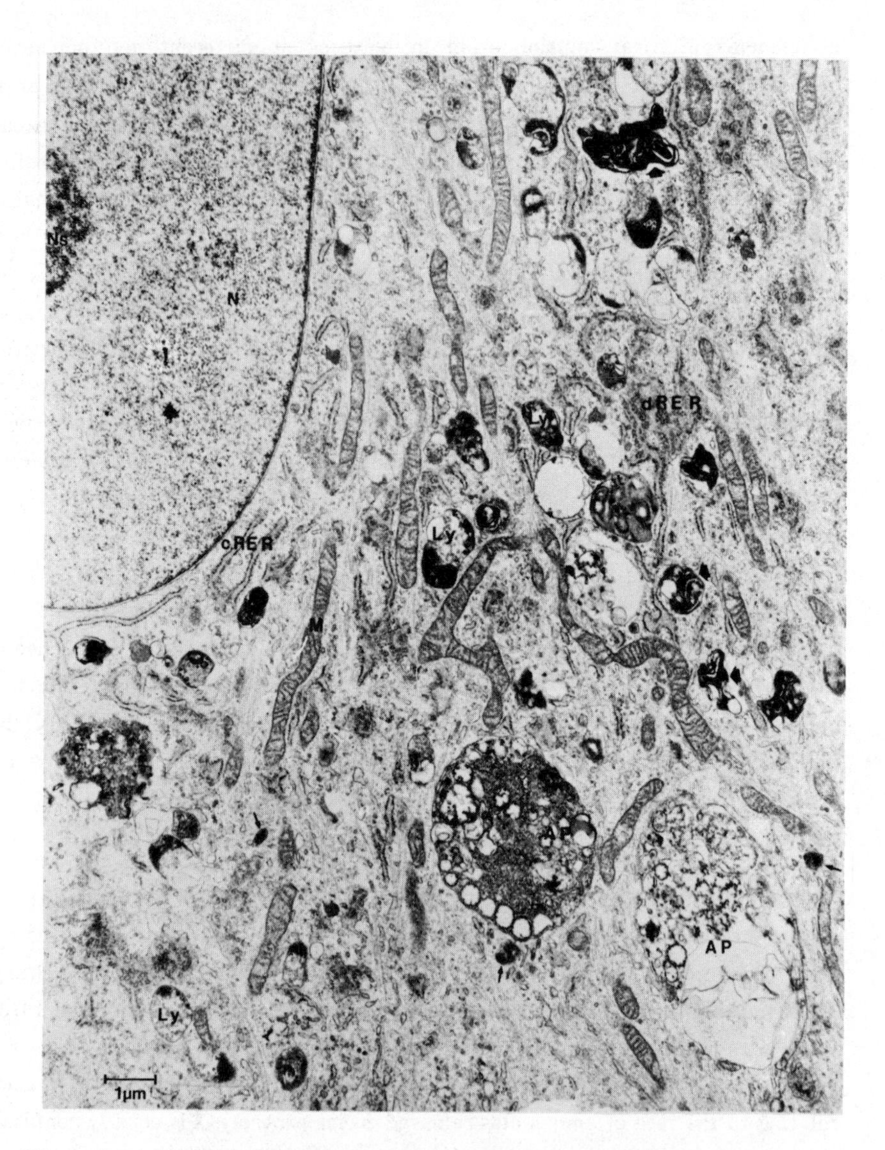

Figure 1: Nuclear chromatin is finely dispersed and there is a single nucleolus (Ns) within the large nucleus (N). Numerous cytoplasmic vacuoles are seen representing various stages of lysosomal development: primary lysosomes (thin arrows), secondary lysosomes (Ly), autophagosomes (AP) and residual myelin-like bodies (thick arrows). The rough endoplasmic reticulum is poorly developed, presenting areas of constriction (cRER) and dilation of the cisternae (dRER). The mitochondria are elongated with transverse and tubular cristae.

efficiencies of viral multiplication in cultures at different cell population doubling (CPD) levels (Holland et al, 1973; Pitha et al, 1974; Tomkins et al, 1974). In a number of instances both theories have been evoked to explain essentially similar results. Thus, while the age-related increase in heat-lability of glucose 6-phosphate dehydrogenase (G6PD) was cited as faulty information transfer between macromolecules (Holliday & Tarrant, 1972), it also has been identified in cultured liver cells as evidence of post-translational modification (Kahn et al, 1977a; Dreyfus et al, 1978). To make the matter even more complicated, it was claimed that increased heat-lability was not seen in purified G6PD from old fibroblasts and therefore that the phenomenon was regulated by the cell medium rather than being an inherent characteristic of the enzyme molecules (Kahn et al, 1977b); but this study was itself subsequently, explicitly criticised on the basis of the experimental protocol adopted (Holliday & Thompson, 1983).

Whatever the mechanisms responsible, it is unlikely that relatively small changes in the specific activity or heat-lability of enzymes such as G6PD are of major consequence in the overall functioning of an individual cell and hence of the culture as a whole. Rather, it might be predicted that the effects of cellular ageing would be manifested in a more primary manner. By this definition, two areas in which investigation would appear warranted are the processes of cellular protein anabolism/catabolism and the utilisation of major nutrient sources in the provision of energy for the growing cultures.

PROTEIN ANABOLISM/CATABOLISM IN AGEING DIPLOID FIBROBLASTS

The rate of protein catabolism in human fibroblasts has been reported both to decline with ageing *in vitro* (Goldstein et al, 1976; Dice, 1982; Gracy, 1983) and to increase (Bradley et al, 1975, 1976; Shakespeare & Buchanan, 1976), the contradictory results obtained probably reflecting the different bases and assumptions of the experimental protocols followed (Makrides, 1983). Evidence relating to the fate of aminoacids released during proteolysis is equally confused. It has been proposed that the intracellular free aminoacid pool occupies a central position and acts directly as a donor and acceptor of aminoacids for anabolic and catabolic processes (Munro, 1970). However a much more complex model incorporating series of catabolic and anabolic pools each with limited exchange facilities also has been considered (Bienkowski & Baum, 1983). In this latter case, the intracellular aminoacid pool would play only a relatively minor part in protein anabolism (Wheatley & Inglis, 1980).

Table 3. Age-related changes in the intracellular free aminoacid pool

		Transport system
Significant decrease*:	Tyrosine	L
	Phenylalanine	L
	Leucine	L
	Isoleucine	L
	Valine	L
Significant increase*:	Methionine	L and A

* One-way analysis of variance, $p < 0.01$

Whether the role of the intracellular aminoacid pool in protein catabolism/anabolism is primary or secondary, an alteration in its composition is likely to be of major importance to the cell. A longitudinal study was established to determine if age-related changes could be identified in the free aminoacid pool of human fibroblasts, with sampling of the cells early in their *in vitro* lifespan, in "middle age" and in the later stages of culture (Sambuy & Bittles, 1982). Besides changes in the intracellular free aminoacid pool secondary to the growth state of the culture (Sambuy & Bittles, 1983), the concentrations of a number of aromatic, branched-chain and sulphur aminoacids showed small but significant increases or decreases that appeared to be specifically related to ageing (Table 3). In view of the considerable emphasis that has been placed on the possible role of specific enzymes in the ageing process(es) of cultured cells, cellular exopeptidases were examined for evidence of changes in specific activity, heat-lability and their activity profiles against a wide range of di- and tripeptides (Sambuy & Bittles, 1984). Despite a small decline in exopeptidase specific activity in the "old" cells and some evidence suggesting greater, overall heat-lability with ageing, the observed, specific, intracellular aminoacid pool changes could not be explained in terms of age-related, altered exopeptidase activity or stability.

The intracellular aminoacid changes could be interpreted as representing reduced availability of tRNA species and/or tRNA synthetases during the life-span of the cells, which in turn would lead to a loss of ability in translating portions of the genetic code (Strehler et al, 1971; Strehler, 1977). A reduction in the specific activity of phenylalanyl tRNA synthetase has been reported in human fibroblasts (Goldstein & Varmuza, 1978) but the data showed a large degree of statistical variance and other workers have failed to confirm a decline in the translational fidelity of old, cultured cells (Harley et al, 1980; Wojtyk &

Goldstein, 1980). In the light of the electron microscopy findings relating to membrane structure alterations with ageing, previously considered, an alternative explanation for the observed changes may be reduced or defective operation of the membrane transport systems responsible for the accumulation of aminoacids within the intracellular pool (Guidotti et al, 1978; Gazzola et al, 1981), in particular the L-system. Regulation of such systems is effected via complex series of feed-back mechanisms involving both intra- and extra-cellular amino-acid concentrations and the implications of an age-dependent decline in the active transport of growth-limiting nutrients, such as the aminoacids, clearly are considerable (Sullivan & De Busk, 1974). Nevertheless, there is little evidence to suggest that either decreased activity of one or more aminoacyl synthetases or a decline in aminoacid active transport would be primary factors in cellular ageing. Clues as to the nature of the basic, causative mechanism(s) must therefore be sought by reference to the uptake and mode of utilisation of major essential energy sources by the fibroblasts.

ENERGY PRODUCTION BY HUMAN DIPLOID FIBROBLASTS

The energy requirements of human diploid fibroblasts *in vitro* are met almost entirely by glucose and glutamine (Sumbilla et al, 1981), the relative utilisation of each compound being subject to reciprocal regulation (Zielke et al, 1978). Under standard growth conditions, approximately one-third of cellular energy needs are provided by the oxidation of glutamine (Zielke et al, 1984) which appears to be the major role of the aminoacid *in vitro* (Griffiths, 1970). With glucose the situation is somewhat different. In cells under normal growth conditions up to 40% of the glucose utilized goes towards energy production, almost entirely via the rapid mechanism of glycolysis (Cristofalo & Kritchevsky, 1966). However, the main function of glucose is to act as a source of ribose moieties for nucleic acid biosynthesis (Zielke et al, 1984).

Recently it has been shown that with ageing there is a marked, biphasic increase in glucose uptake from the growth medium that can be accounted for in terms of lactate output and production (Fig. 2). This occurs prior to any change in intracellular glucose or lactate levels or in the specific activities of enzymes of the glycolytic pathway (Bittles & Harper, 1984). It previously had been observed that, although the uptake of glutamine from the growth medium declined by up to 27% when "young" and "old" cells were compared (Sambuy, 1982), there was a concomitant, significant increase in the intracellular glutamine concentration (Sambuy & Bittles, 1982). Considered together, these

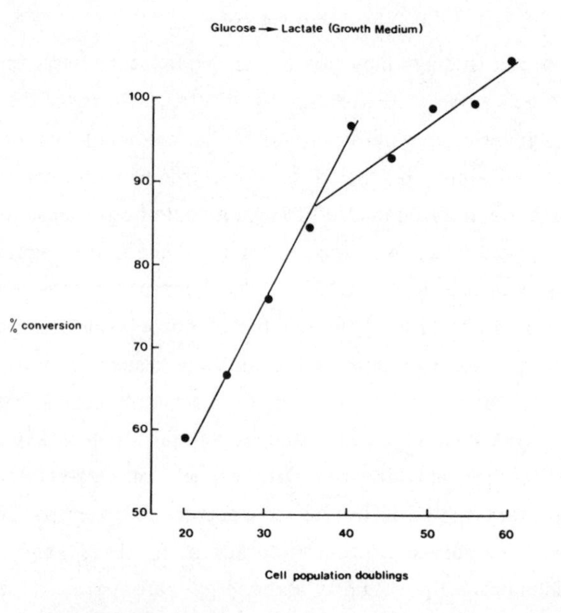

Figure 2: Percentage uptake of glucose from the growth medium accounted for in terms of lactate production, measured at each fifth cell population doubling (CPD). (Reproduced by kind permission of the Biochemical Society.)

two sets of results suggest a gradual decline in glutamine oxidation with ageing, leading to a potential shortfall in cellular energy production that is made good via glycolysis. Any shift in glucose utilisation by the cells from a predominantly anabolic mode to glycolysis has major implications, as it necessarily must reduce their ability to maintain nucleic acid biosynthesis.

It is noteworthy that the major switch to glycolysis by CPD 40 (Fig. 2) corresponds very closely to the proposed transition from phase II of growth to phase III (Macieira-Coelho & Taboury, 1982), characterised by a decrease in the number of cells dividing at each passage and accompanied by the onset of changes in cellular nucleoprotein organisation (Puvion-Dutilleul et al, 1982). The further, lesser increase in glycolysis between CPD 45 and the end of cell division by CPD 60 almost certainly indicates that once a critical point is exceeded and the growth phase II/III transition completed, the culture is irreversibly committed to the eventual cessation of cell multiplication, seen as phase IV. However, it is worth emphasizing that the entry of the culture into phase IV does not preclude continued cell maintenance in a viable, non-dividing state for considerable periods of time (Cutler, 1982).

CONCLUSIONS

The glycolysis findings may provide an explanatory basis for many of the observed phenomena associated with ageing *in vitro*. However the nature of the factor(s) which governs the putative switch in the mode of cellular energy production from glutamine oxidation to glycolysis remains unidentified. An early, gradual decline in mitochondrial function could be responsible and, in fact, many reports have claimed major mitochondrial changes with ageing *in vivo* and *in vitro* (reviewed by Hansford, 1981). The general changes most commonly observed have been a decrease in the number of mitochondria and an increase in their average size: for example, in human liver (Tauchi & Sato, 1968), in rat heart (Levkova & Trunov, 1970) and even in the flight muscle mitochondria of blowflies (Tribe & Ashhurst, 1972). Recent animal studies have improved the levels of understanding in this area with reports on age-related declines in mitochondrial protein synthesis (Bailey & Webster, 1984), in particular the biosynthesis of inner membrane proteins (Marcus et al, 1982), and a decrease in mitochondrial membrane lipid fluidity and energy transduction (Lewin & Timiras, 1984). Significantly, a reduction in oxidative phosphorylation, ATP synthesis and APTase activity associated with lipid changes in the inner-membrane matrix compartment of rat heart mitochondria has been noted by four months of age (Clandinin & Innis, 1983). Changes of this nature would appear to indicate that the primary cellular effect in ageing may be a conformational alteration of mitochondrial inner membrane structure leading to uncoupling of oxidative phosphorylation, the consequent effect bearing maximally on state 3 respiration.

Although our preliminary studies on human diploid fibroblasts as yet have been unable to confirm uncoupling of oxidative phosphorylation, the observed shift to glycolysis does appear to complement and to extend the theoretical, evolutionary interpretation of ageing advanced by Kirkwood & Holliday (this volume). Thus, while ageing may indeed represent a price to be paid for energy-saving in somatic cells (Kirkwood, 1977), the attempted maintenance of energy production via glycolysis, at the expense of nucleic acid biosynthesis, would appear an equally critical factor. Whether the patterns of change in energy source utilisation further can be reconciled with age-related changes in gene expression, such as reduced methylation (Shmookler Reis & Goldstein, 1980; Wilson & Jones, 1983) remains a subject for speculation and future study.

REFERENCES

Aizawa, S., Mitsui, Y., Kurimoto, F. & Nomura, K. (1980). Cell surface changes accompanying aging in human diploid fibroblasts. V. Role of large major cell surface protein and surface negative charge in aging- and transformation-associated changes in Concanavalin A- mediated red blood cell adsorption. Experimental Cell Research, **127**, 143-157.

Bailey, P.J. & Webster, G.C. (1984). Lowered rates of protein synthesis by mitochondria isolated from organisms of increasing age. Mechanisms of Ageing and Development, **24**, 233-241.

Basler, J.W., David, J.D. & Agris, P.F. (1979). Deteriorating collagen synthesis and cell ultrastructure accompanying senescence of human normal and Werner's syndrome fibroblast cell strains. Experimental Cell Research, **118**, 73-84.

Bell, E., Marek, L.F., Levinstone, D.S., Merrill, C., Sher, S., Young, I.T. & Eden, M. (1978). Loss of division potential *in vitro*: aging or differentiation? Science, **202**, 1158-1163.

Bemiller, P.M. & Lee, L. (1976). Nucleolar changes in senescing WI-38 cells. Mechanisms of Ageing and Development, **8**, 417-427.

Bemiller, P.M. & Miller, J.E. (1979). Cytological changes in senescing WI-38 cells: a statistical analysis. Mechanisms of Ageing and Development, **10**, 1-5.

Bienkowski, R.S. & Baum, B.J. (1983). Measurement of intracellular protein degradation. In: R.C. Adelman & G.S. Roth (eds.), Altered Proteins and Aging, p. 55 Boca Raton, Florida: CRC Press.

Bierman, E.L. (1978). The effect of donor age on the *in vitro* life span of cultured human arterial smooth muscle cells. In Vitro, **14**, 951-955.

Bittles, A.H. & Harper, N. (1984). Increased glycolysis in ageing cultured human diploid fibroblasts. Bioscience Reports, **4**, 751-756.

Boak, A.M., Bittles, A.H. & Quinn, P.J. (1982). Age-related ultrastructural changes in human embryonic lung fibroblasts. Experimental Gerontology, **18**, 139-146.

Bosmann, H.B., Gutheil, R.L. & Case, K.R. (1975). Loss of a critical neutral protease in ageing WI-38 cells. Nature, **261**, 499-501.

Bowman, P.D., Meek, R.L. & Daniel, C.W. (1975). Aging of human fibroblasts *in vitro*. Correlations between DNA synthetic ability and cell size. Experimental Cell Research, **93**, 184-190.

Bradley, M.O., Dice, J.F., Hayflick, L. & Schimke, R.T. (1975). Protein alterations in aging WI-38 cells as determined by proteolytic susceptibility. Experimental Cell Research, **96**, 103-112.

Bradley, M.O., Hayflick, L. & Schimke, R.T. (1976). Protein degradation in human fibroblasts (WI-38). Effects of aging, viral transformation and aminoacid analogs. Journal of Biological Chemistry, **251**, 3521-3529.

Carrel, A. (1912). On the permanent life of tissues outside of the organism. Journal of Experimental Medicine, **15**, 516-528.

Carrel, A. (1914). Present condition of a strain of connective tissue twenty-eight months old. Journal of Experimental Medicine, **20**, 1-2.

Carrel, A. & Burrows, M.T. (1910). Cultivation of adult tissues and organs outside of the body. Journal of the American Medical Association, **55**, 1379-1381.

Clandinin, M.T. & Innis, S.M. (1983). Does mitochondrial ATP synthesis decline as a function of change in the membrane environment with ageing? Mechanisms of Ageing and Development, **22**, 205-208.

Cristofalo, V.J. & Kritchevsky, D. (1966). Respiration and glycolysis in the human diploid cell strain WI-38. Journal of Cellular Physiology, **67**, 125-132.

Cutler, R.G. (1982). Longevity is determined by specific genes: testing the hypothesis. In: R.C. Adelman & G.S. Roth (eds.), Testing the Theories of Aging, p. 26. Boca Raton, Florida: CRC Press.

Dell'Orco, R.T., Mertens, J.G. & Kruse, P.F. (1973). Doubling potential, calendar time, and senescence of human diploid cells in culture. Experimental Cell Research, **77**, 356-360.

Dice, J.F. (1982). Altered degradation of proteins microinjected into senescent human fibroblasts. Journal of Biological Chemistry, **257**, 14624-14627.

Dreyfus, J.-C., Kahn, A. & Schapira, F. (1978). Post-translational modifications of enzymes. Current Topics in Cellular Regulation, **14**, 243-297.

Dreyfus, J.-C., Kahn, A. & Schapira, F. (1983). Molecular mechanisms of alterations of some enzymes in aging. In: R.C. Adelman & G.S. Roth (eds.), Altered Proteins and Aging, p. 113. Boca Raton, Florida: CRC Press.

Ebeling, A.H. (1922). A ten year old strain of fibroblasts. Journal of Experimental Medicine, **35**, 755-759.

Gazzola, G.C., Dall'Asta, V. & Guidotti, G.G. (1980). The transport of neutral aminoacids in cultured human fibroblasts. Journal of Biological Chemistry, **255**, 929-936.

Gazzola, G.C., Dall'Asta, V. & Guidotti, G.G. (1981). Adaptive regulation of aminoacid transport in cultured human fibroblasts. Sites and mechanisms of action. Journal of Biological Chemistry, **256**, 3191-3198.

Gershon, D. (1979). Current status of age-altered enzymes: alternative mechanisms. Mechanisms of Ageing and Development, **9**, 189-196.

Gershon, D. & Gershon, H. (1976). An evaluation of the "error catastrophe" theory of ageing in the light of recent experimental results. Gerontology, **22**, 212-219.

Gilchrest, B.A. (1983). In vitro assessment of keratinocyte aging. Journal of Investigative Dermatology, **81**, 184s-189s.

Goldstein, S. (1969). Lifespan of cultured cells in progeria. Lancet **i**, 424.

Goldstein, S. & Singal, D.P. (1974). Senesence of cultured human fibroblasts: mitotic versus metabolic time. Experimental Cell Research, **88**, 359-364.

Goldstein, S. & Varmuza, S.L. (1978). Phenylalanyl synthetase function in cultured fibroblasts from subjects with progeria. Canadian Journal of Biochemistry, **56**, 73-79.

Goldstein, S., Stotland, D. & Cordeiro, R.A.J. (1976). Decreased proteolysis and increased aminoacid efflux in aging human fibroblasts. Mechanisms of Ageing and Development, **5**, 221-233.

Goldstein, S., Moerman, E.J., Soeldner, J.S., Gleason, R.E. & Barnett, J.M. (1979). Diabetes mellitus and genetic prediabetes. Decreased replicative capacity of cultured skin fibroblasts. Journal of Clinical Investigation, **63**, 358-370.

Gracy, R.W. (1983). Epigenetic formation of isozymes; the effect of aging. Isozymes: Current Topics in Biology and Medical Research, **7**, 187-201.

Gracy, R.W., Lu, H.S., Yuan, P.M. & Talent, J.M. (1983). Structural analysis of altered proteins. In: R.C. Adelman & G.S. Roth (eds.), Altered Proteins and Aging, p. 9. Boca Raton, Florida: CRC Press.

Griffiths, J.B. (1970). The quantitative utilization of aminoacids and glucose and contact inhibition of growth in cultures of the human diploid cell, WI-38. Journal of Cell Science, **6**, 739-749.

Greenberg, S.B., Grove, G.L. & Cristofalo, V.J. (1977). Cell size in aging monolayer cultures. In Vitro, **13**, 297-300.

Guidotti, G.G., Borghetti, A.F. & Gazzola, G.C. (1978). The regulation of aminoacid transport in animal cells. Biochimica et Biophysica Acta, **515**, 329-366.

Hansford, R.G. (1981). Energy metabolism. In: J.R. Florini (ed.), Handbook of Biochemistry in Aging, p. 137. Boca Raton, Florida: CRC Press.

Harley, C.B. & Goldstein, S. (1978). Cultured human fibroblasts: distribution of cell generations and a critical limit. Journal of Cellular Physiology, **97**, 509-516.

Harley, C.B., Pollard, J.W., Chamberlain, J.W., Stanners, C.P. & Goldstein, S. (1980). Protein synthetic errors do not increase during aging of cultured human fibroblasts. Proceedings of the National Academy of Sciences, USA., **77**, 1885-1889.

Harrison, R.G. (1907). Observations on the living developing nerve fiber. Proceedings of the Society for Experimental Biology and Medicine, **4**, 140-143.

Hayflick, L. (1965). The limited *in vitro* lifetime of human diploid cell strains. Experimental Cell Research, **37**, 614-636.

Hayflick, L. (1980a). Recent advances in the cell biology of aging. Mechanisms of Ageing and Development. **14**, 59-79.

Hayflick, L. (1980b). Cell aging. Annual Review of Gerontology and Geriatrics, **1**, 26-67.

Hayflick, L. & Moorhead, P.S. (1961). The serial cultivation of human diploid cell strains. Experimental Cell Research, **25**, 585-621.

Hirsch, G.P. (1982). Error theories and fidelity in aging and cancer. In: R.C. Adelman & G.S. Roth (eds.), Testing the Theories of Aging, p. 184. Boca Raton, Florida: CRC Press.

Holland, J.J., Kohne, D. & Doyle, M.V. (1973). Analysis of virus replication in ageing human fibroblast cultures. Nature, **245**, 316-318.

Holliday, R. & Tarrant, G.M. (1972). Altered enzymes in ageing human fibroblasts. Nature, **238**, 26-30.

Holliday, R. & Thompson, K.V.A. (1983). Genetic effects on the longevity of cultured human fibroblasts. III. Correlations with altered glucose-6-phosphate dehydrogenase. Gerontology, **29**, 89-96.

Holliday, R., Porterfield, J.S. & Gibbs, D.D. (1974). Premature ageing and occurrence of altered enzyme in Werner's syndrome fibroblasts. Nature, **248**, 762-763.

Holliday, R., Huschtscha, L.I., Tarrant, G.M. & Kirkwood, T.B. (1977). Testing the commitment theory of cellular aging. Science, **198**, 366-372.

Johnson, J.E. (1979). Fine structure of IMR-90 cells in culture as examined by scanning and transmission electron microscopy. Mechanisms of Ageing and Development, **10**, 405-443.

Johnson, J.E. (1984). In vivo and in vitro comparisons of age-related fine structural changes in cell components. In: J.E. Johnson (ed.), Aging and Cell Structure, vol. 2, p. 37. New York: Plenum Press.

Kahn, A., Guillozo, A., Cottreau, D., Marie, J., Bourel, M., Boivin, P. & Dreyfus, J.-C. (1977a). Accuracy of protein synthesis and *in vitro* ageing: search for altered enzymes in senescent cultured cells from human livers. Gerontology, **23**, 174-184.

Kahn, A., Guillozo, A., Leibovitch, M.-P., Cottreau, D., Bourel, M. & Dreyfus, J.-C. (1977b). Heat-lability of glucose-6-phosphate dehydrogenase in some senescent human cultured cells. Evidence for its post-synthetic nature. Biochemical and Biophysical Research Communications, **77**, 760-766.

Kelley, R.O. (1976). Development of the aging cell surface: a freeze-fracture analysis of gap junctions between human embryo fibroblasts aging in culture. A brief note. Mechanisms of Ageing and Development, **5**, 339-345.

Kelley, R.O. & Skipper, B.E. (1977). Development of the aging cell surface: variation in the distribution of intramembrane particles with progressive age of human diploid fibroblasts. Journal of Ultrastructure Research, **59**, 114-118.

Kelley, R.O. & Vogel, K.G. (1984). The aging cell surface: structural and biochemical alterations associated with progressive sub-cultivation of human diploid fibroblasts. In: J.E. Johnson (ed.), Aging and Cell Structure, vol. 2, p. 2. New York: Plenum Press.

Kelley, R.O., Trotter, J.A., Marek, L.F., Perdue, B.D. & Taylor, C.B. (1980). Development of the aging cell surface: variation in cytoskeletal assembly during spreading of progressively subcultivated human embryo fibroblasts (IMR-90). Mechanisms of Ageing and Development, **13**, 127-141.

Kirkwood, T.B.L. (1977). Evolution of ageing. Nature, **270**, 301-304.

Kirkwood, T.B.L. & Holliday, R. (1975). Commitment to senescence: a model for the finite and infinite growth of diploid and transformed human fibroblasts in culture. Journal of Theoretical Biology, **53**, 481-496.

Kohn, R.R. (1982). Evidence against cellular aging theories. In: R.C. Adelman & G.S. Roth (eds.), Testing the Theories of Aging, p. 222. Boca Raton, Florida: CRC Press.

Lee, S.-C., Bemiller, P.M., Bemiller, J.N. & Pappelis, A.J. (1978). Nuclear area changes in senescing human diploid fibroblasts. Mechanisms of Ageing and Development, **7**, 417-424.

Le Guilly, Y., Simon, M., Lenoir, P. & Bourel, M. (1973). Long-term culture of human adult liver cells; morphological changes related to *in vitro* senescence and effect of donor's age on growth potential. Gerontologia, **19**, 303-313.

Levkova, H.A. & Trunov, B.H. (1970). Features peculiar to the structure of mitochondria of the senile myocardium. Kardiologiia, **10**, 94-97.

Lewin, M.B. & Timiras, P.S. (1984). Lipid changes with aging in cardiac mitochondrial membranes. Mechanisms of Ageing and Development, **24**, 343-352.

Lipetz, J. & Cristofalo, V.J. (1982). Ultrastructural changes accompanying the aging of human diploid cells in culture. Journal of Ultrastructural Research, **39**, 43-56.

Macieira-Coelho, A. & Taboury, F. (1982). A re-evaluation of the changes in proliferation in human fibroblasts during ageing *in vitro*. Cell Tissue Kinetics, **15**, 213-224.

Macieira-Coelho, A., Ponten, J. & Philipson, L. (1966a). The division cycle and RNA-synthesis in diploid human cells at different passage levels *in vitro*. Experimental Cell Research, **42**, 673-684.

Macieira-Coelho, A., Ponten, J. & Philipson, L. (1966b). Inhibition of the division cycle in confluent cultures of human fibroblasts *in vitro*. Experimental Cell Research, **43**, 20-29.

Makrides, S.C. (1983). Protein synthesis and degradation during aging and senescence. Biological Review of the Cambridge Philosophical Society, **58**, 353-422.

Marcus, D.L., Ibrahim, N.G. & Freedman, M.L. (1982). Age-related decline in the biosynthesis of mitochondrial inner membrane proteins. Experimental Gerontology, **17**, 333-341.

Martin, G.M., Sprague, C.A. & Epstein, C.J. (1970). Replicative life-span of cultivated human cells. Effects of donor's age, tissue and genotype. Laboratory Investigation, **23**, 86-92.

Miquel, J. & Fleming, J.E. (1984). A two-step hypothesis on the mechanisms of in vitro cell aging: cell differentiation followed by intrinsic mitochondrial mutagenesis. Experimental Gerontology, **19**, 31-36.

Mitsui, Y. & Schneider, E.L. (1976a). Characterization of fractionated human diploid fibroblast cell populations. Experimental Cell Research, **103**, 23-30.

Mitsui, Y. & Schneider, E.L. (1976b). Relationship between cell replication and volume in senescent human diploid fibroblasts. Mechanisms of Ageing and Development, **5**, 45-56.

Mitsui, Y., Matsuoka, K., Aizawa, S. & Noda, K. (1980). New approaches to characterization of aging human fibroblasts at individual cell level. Advances in Experimental Biology and Medicine, **129**, 5-25.

Munro, H.N. (1970). Free aminoacid pools. In: H.N. Munro (ed.), Mammalian Protein Metabolism, vol. 4, p. 299. New York: Academic Press.

Orgel, L.E. (1963). The maintenance of the accuracy of protein synthesis and its relevance to ageing. Proceedings of the National Academy of Sciences, USA., **49**, 517-521.

Orgel, L.E. (1970). The maintenance of the accuracy of protein synthesis and its relevance to ageing: a correction. Proceedings of the National Academy of Sciences, USA, **67**, 1476.

Orgel, L. (1973). Ageing of clones of mammalian cells. Nature, **243**, 441-445.

Pieraggi, M.T., Julian, M. & Bouissou, H. (1984). Fibroblast changes in cutaneous ageing. Virchows Archiv. A, **402**, 275-287.

Pitha, J., Adams, R. & Pitha, P.M. (1974). Viral probe into the events of cellular (*in vitro*) aging. Journal of Cellular Physiology, **83**, 211-218.

Ponten, J., Stein, W.D. & Shall, S. (1983). A quantitative analysis of the aging of human glial cells in culture. Journal of Cellular Physiology, **117**, 342-352.

Pool, T.B. & Metter, J.D. (1984). New concepts in regulation of the lifespan of human diploid fibroblasts *in vitro*. In: J.E. Johnson (ed.), Aging and Cell Structure, vol. 2, p. 89. New York: Plenum Press.

Puvion-Dutilleul, F., Azzarone, B. & Macieira-Coelho, A. (1982). Comparison between proliferative changes and nuclear events during ageing of human fibroblasts *in vitro*. Mechanisms of Ageing and Development, **20**, 75-92.

Reiner, J.M. (1983). Differentiation, ageing and terminal differentiation: a semantic analysis. Journal of Theoretical Biology, **105**, 545-552.

Rheinwald, J.G. & Green, H. (1975). Serial cultivation of strains of human epidermal keratinocytes: the formation of keratinizing colonies from single cells. Cell, **6**, 331-344.

Robert, L. (1977). Membranes and ageing. In: G.A. Jamieson & D.M. Robinson (eds.), Mammalian Cell Membranes, vol. 5, Responses of the Plasma Membrane, p. 220. London: Butterworth.

Robbins, E., Levine, E.M. & Eagle, H. (1970). Morphologic changes accompanying senescence of cultured human diploid cells. Journal of Experimental Medicine, **131**, 1211-1222.

Rothstein, M. (1977). Recent developments in the age-related alteration of enzymes: a review. Mechanisms of Ageing and Development, **6**, 241-257.

Roux, W. (1885). Beitrage zur entwicklungsmechanik des embryo. Zeitschrift fur Biologie, **21**, 411-526.

Sambuy, Y. (1982). The effects of ageing processes on the free aminoacid pool of cultured human fibroblasts. Ph.D. thesis, University of London.

Sambuy, Y. & Bittles, A.H. (1982). The effects of *in vitro* ageing on the composition of the intracellular free aminoacid pool of human diploid fibroblasts. Mechanisms of Ageing and Development, **20**, 279-287.

Sambuy, Y. & Bittles, A.H. (1983). Growth-dependent and age-related changes in the free aminoacid pool of human diploid fibroblasts. In: A. Castellani (ed.), The Use of Human Cells for the Evaluation of Risk from Physical and Chemical Agents, p. 735. New York: Plenum Press.

Sambuy, Y. & Bittles, A.H. (1984). The effects of *in vitro* ageing on the exopeptidases of human diploid fibroblasts. Mechanisms of Ageing and Development, **26**, 13-22.

Schneider, E.L. & Fowlkes, B.J. (1976). Measurement of DNA content and cell volume in senescent human fibroblasts utilizing flow multiparameter single cell analysis. Experimental Cell Research, **98**, 298-302.

Schneider, E.L. & Mitsui, Y. (1976). The relationship between *in vitro* cellular ageing and *in vivo* human age. Proceedings of the National Academy of Sciences, USA, **73**, 3584-3588.

Shakespeare, V. & Buchanan, J.H. (1976). Increased degradation rates of protein in ageing human fibroblasts and in cells treated with an aminoacid analog. Experimental Cell Research, **100**, 1-8.

Shapiro, B.L., Lam, L.F. & Fast, L.H. (1979). Premature senescence in cultured skin fibroblasts from subjects with cystic fibrosis. Science, **203**, 1251-1253.

Shmookler Reis, R.J. & Goldstein, S. (1980). Loss of reiterated DNA sequences during serial passage of human diploid fibroblasts. Cell, **21**, 739-749.

Shock, N.W. (1981). Biological theories of aging. In: J.R. Florini (ed.), Handbook of Biochemistry in Aging, p. 271. Boca Raton, Florida: CRC Press.

Simons, J.W.I.M. & van der Broeck, C. (1970). Comparison of ageing *in vitro* and ageing *in vivo* by means of cell size analysis by using a Coulter counter. Gerontologia, **16**, 340-351.

Smith, J.R., Pereira-Smith, O.M. & Schneider, E.L. (1978). Colony size distributions as a measure of *in vivo* and *in vitro* aging. Proceedings of the National Academy of Sciences, USA, **75**, 1353-1356.

Strehler, B.L. (1977). Time, Cells and Aging, 2nd edn, p. 307. New York: Academic Press.

Strehler, B.L., Birsch, G., Gussek, D., Johnson, R. & Bick, M. (1971). Codon-restriction theory of aging and development. Journal of Theoretical Biology, **33**, 429-474.

Sullivan, J.L. & De Busk, A.G. (1974). Membrane glycoproteins and cellular aging. Journal of Theoretical Biology, **46**, 291-294.

Sumbilla, C.M., Zielke, C.L., Reed, W.E., Ozand, P.T. & Zielke, H.R. (1981). Comparison of the oxidation of glutamine, glucose, ketone bodies and fatty acids by human diploid fibroblasts. Biochimica et Biophysica Acta, **675**, 301-304.

Tauchi, H. & Sato, T. (1968). Age changes in size and number of mitochondria of human hepatic cells. Journal of Gerontology, **23**, 454-461.

Thaw, H.H., Collins, V.P. & Brunk, U.T. (1984). Influence of oxygen tension, pro-oxidants and anti-oxidants on the formation of lipid peroxidation products (lipofuscin) in individual cultured human glial cells. Mechanisms of Ageing and Development, **24**, 211-224.

Tompkins, G.A., Stanbridge, E.J. & Hayflick, L. (1974). Viral probes of aging in the human diploid cell strain WI-38. Proceedings of the Society for Experimental Biology and Medicine, **146**, 385-390.

Tribe, M.A. & Ashhurst, D.E. (1972). Biochemical and structural variation in the flight muscle mitochondria of aging blowflies, *Calliphora erythrocephala*. Journal of Cell Science, **10**, 443-469.

Vogel, K.G., Kelley, R.O. & Stewart, C. (1981). Loss of organized fibronectin matrix from the surface of aging diploid fibroblasts (IMR-90). Mechanisms of Ageing and Development, **16**, 295-302.

Vracko, R., McFarland, B.H. & Pecoraro, R.E. (1983). Seeding efficiency, plating efficiency and population doublings of human skin fibroblast-like cells: results of replicative testing. In Vitro, **19**, 504-514.

Walton, J. (1982). The role of limited cell replicative capacity in pathological age change. A review. Mechanisms of Ageing and Development, **19**, 217-244.

Waters, H. & Walford, R.L. (1970). Latent period for outgrowth of human skin explants as a function of age. Journal of Gerontology, **25**, 381-383.

Wen, W.N., Liew, T.-L., Wuu, S.W. & Jan, K.Y. (1983). The effect of age and cell proliferation on the frequency of sister chromatid exchange in human lymphocytes cultured *in vitro*. Mechanisms of Ageing and Development, **21**, 377-384.

Wheatley, D.N. & Inglis, M.S. (1980). An intracellular perfusion system linking pools and protein synthesis. Journal of Theoretical Biology, **83**, 437-445.

Wilson, V.L. & Jones, P.A. (1983). DNA methylation decreases in aging but not in immortal cells. Science, **220**, 1055-1057.

Wojtyk, R.I. & Goldstein, S. (1980). Fidelity of protein synthesis does not decline during aging of cultured human fibroblasts. Journal of Cellular Physiology, **103**, 299-303.

Wolosewick, J.J. & Porter, K.R. (1977). Observations on the morphological heterogeneity of WI-38 cells. American Journal of Anatomy, **149**, 197-226.

Zielke, H.R., Ozand, P.T., Tildon, J.T., Sevdalian, D.A. & Cornblath, M. (1978). Reciprocal regulation of glucose and glutamine utilization by cultured human diploid fibroblasts. Journal of Cellular Physiology, **95**, 41-48.

Zielke, H.R., Zielke, C.L. & Ozand, P.T. (1984). Glutamine: a major energy source for cultured mammalian cells. Federation Proceedings, **43**, 121-125.

ESTIMATION OF BIOLOGICAL MATURITY IN THE OLDER CHILD

M. A. PREECE and L. A. COX
Department of Growth and Development
Institute of Child Health, London

INTRODUCTION

It is well recognised that chronological age (CA) is a poor measure of maturity (Tanner, 1962). An average-maturing boy completes his growth in stature at 18 years of age whereas an early developer may do so at 15 or 16 years; the late-developer will still be growing as late as 20 years of age. With girls the same is true except that the whole process occurs some two years earlier. In a similar way, an average boy shows the first obvious signs of puberty, genitalia stage 2, at an age of 12 years. In contrast, the earliest developing boy within the normal range will do the same before his tenth birthday and a rather late-developing boy may still show no signs of puberty at the fourteenth birthday. In girls similar situations obtain: the average girl reaches breast stage 2 soon after 11 years of age but the earliest developers may show the same signs by nine years of age and the latest not before the thirteenth birthday. These variations in the progress of maturation introduce a new dimension to the process of growth and development, for which the name 'tempo' was coined by Boas (Tanner, 1981). Quantification of tempo requires the use of some measure of maturational progress and it is therefore helpful to explore what methods are available and compare them in terms of their value and validity.

At present there are four basic maturity indices which are in fairly regular use. These are assessments of: pubertal stage; bone age (BA); dental age; and height age. The first three are perfectly valid measures although sometimes rather inadequate to the job they are asked to perform; the fourth is completely invalid, as will be discussed below.

MATURITY INDICES

Pubertal Stages

If the assessment of maturity is limited to the peripubertal stage of growth and development, then the stages of puberty provide a convenient and

reliable assessment of maturation. Almost universally the systems in current use depend on the criteria that were formalised by Tanner (1962). These depend upon rating breast and pubic hair development on a five point scale and axillary hair on a three point scale for girls, and external genitalia, pubic and axillary hair on similar scales for boys. These stages are all well described in other sources and will not be further enumerated here. Suffice it to say that in all these systems stage 1 is the prepubertal state and stage 5 (or 3 in the case of axillary hair) indicate full maturity. Conventionally the stages are abbreviated such that, for example, breast stage 2 is referred to as B2, genitalia stage 4 as G4.

There are three major problems with pubertal stages: they are only relevant in the peripubertal child; they are relatively subjective; and by its nature the staging is rather coarse.

Bone Age

There are three current methods employed for assessing skeletal age although there have been earlier methods often contributing to the evolution of newer techniques. The older method still in use is of the atlas type first described in 1950 by Greulich and Pyle and subsequently revised (1959). In this method, a sequence of hand-wrist radiographs typical of children of specific CAs are displayed. While the authors intended the user to compare carefully each bone of the patient's X-ray with the appropriate standard and then derive a median skeletal age, the tendency is simply to find the example that most closely resembles the patient's and call that the bone 'age'. This is very unsatisfactory as the observer easily becomes focused on a few bones, especially the radius, ulna and carpus, leading to a biased assessment. A further problem is that the Greulich and Pyle method was based mainly on the radiographs of the Brush Foundation Study (Todd, 1937), which consisted of socially advantaged children. They thus had relatively advanced skeletal maturation compared to other American children and even more so to European children. At the present time the majority of North American children are still between six and twelve months delayed compared to the Greulich and Pyle standards as are all average maturing European children. Other systems for assessing maturation, similar in style to that of Greulich and Pyle but having the same limitations, are based on the knee or foot and ankle (Pyle & Hoerr, 1955).

Preferred methods use the bone-specific scoring systems for the hand-wrist (TW2) (Tanner et al, 1983) and the knee (Roche et al, 1975). In these

Table 1. Means, standard deviations (SD) and numbers of observations (n) for chronological age (CA) of attaining the various events of puberty.

		BOYS			GIRLS		
		Mean (years)	SD	n	Mean (Years)	SD	n
G/B Stage	2	11.16	1.03	50	10.81	0.89	34
	3	12.63	1.02	50	11.68	1.05	36
	4	13.55	0.91	48	12.48	1.09	26
	5	14.63	0.97	45	13.89	1.30	37
PH Stage	2	12.37	1.46	47	11.30	1.07	29
	3	13.31	0.85	47	11.86	1.01	37
	4	13.99	0.78	45	12.75	1.08	35
	5	15.10	1.05	47	14.07	1.04	34
Menarche					12.98	1.04	39
PHV-age		13.84	0.84		11.65	0.89	39
PSHV-age		14.39	0.83	51	12.35	1.02	39
PSLLV-age		13.08	0.89	51	11.08	0.86	39
Height	MH90	13.74	0.79	51	11.99	0.83	39
	MH95	14.86	1.02	51	12.81	0.80	39
Sitting Height	MH90	14.10	0.74	51	11.99	0.83	39
	MH95	15.14	0.75	51	13.21	0.84	39

G/B genitalia/breast stage
PH pubic hair stage
PHV-age peak height velocity - age
PSHV-age peak sitting height velocity - age
PSLLV-age peak sub-ischial leg length velocity - age
MH90 90% mature height

methods each bone is assessed according to a number of specific maturity indicators to which a score is attached. Only after scoring each indicator is it possible to derive a total maturity score which may then be translated into a bone 'age' by a comparison with appropriate standards. Thus it is impossible to

use a shortcut as in the atlas method. The methods can be used for any population although it is wise to develop specific standards for each country by first scoring X-rays of children of the appropriate background.

Dental Age

The concept of dental age has existed since the 1920s; initially it was based on a count of the number of erupted teeth, deciduous or permanent, but it has more recently been refined in a number of ways. When purely dependent on visible teeth it has the attraction of being immediate, but it is unfortunately very crude. In general it offers little in the present context and will not be considered further. It was reviewed in more detail in Demirjian (1978).

Height Age

The term height age (HA) has permeated much growth disorder literature since its use was suggested by Wilkins (1950). This is unfortunate as it incorporates a basic misunderstanding of the requirements for defining developmental age. An implicit assumption in a developmental age is that all individuals reach the same mature end-point. This is clearly not so for stature. A child may have a delayed HA for two reasons: delayed maturation, an absolute short final stature, or both. A child has a delayed skeletal age for only one reason: delayed maturation. These differences are confounded by HA and may lead to serious misunderstandings.

Height age is assigned by noting the age at which the average child of the same sex has the same height as the patient. It does not recognise that the child's family might be unusually short (or tall) and that therefore the population average is not relevant. Further, it takes no note of the normal variation within a family which is not dependent on maturity and is about 17 cm at adult size.

While HA has no place in modern auxological thinking, it might be attractive to use a variable directly related to physical growth as a measure of maturity. However, the clear inadequacy of HA requires us to look further and this aspect will now be pursued in some depth.

SIZE-RELATED VARIABLES AS MEASURES OF MATURITY

In what follows it will frequently be necessary to refer to a scale of maturation such as BA. When so doing 'age' provides a convenient general term

for all scales, be they CA, BA or size-related as described below. Therefore, 'age' should be interpreted as being merely a measure of the passage of time on any of several scales; which is intended to be clear from the context.

There are really two possibilities in which height data could be used to characterise the tempo of an individual child, but both of these are essentially a retrospective exercise and only of use in long-term longitudinal studies. The measures that seem most useful are both in the form of an age at which a particular landmark of the growth curve is achieved. On the one hand there is the age at which the peak growth velocity of the adolescent growth spurt is attained. Most commonly this will be for height (PHV-age), but there is no reason why this should not be extended to other skeletal measurements. The second method is to look at the age of attaining a particular percentage of mature height. This would usually be a relatively high percentage, somewhat near to the completion of maturity, at say 90 per cent or 95 per cent of mature size.

Both of these measures require the availability of substantial longitudinal growth data and in the past have also required rather extensive graphical methods for the calculation of the various ages for different events. More recently the second limitation has been effectively removed. It is possible to fit mathematical models to the growth data of individual children so that their serial measurements are summarised in a limited number of parameters of a fixed mathematical model (Preece & Heinrich, 1981; Marubini, 1984). There are a number of different models that could be used for this, but for the purposes of the present study that described by Preece & Baines (1978) has been used. This model contains five parameters which are estimated by appropriate non-linear curve-fitting techniques. From these five parameters it is possible to calculate PHV-age and further, the age at which any particular proportion of mature height was achieved. For the present purposes we have studied the age of 90 per cent (MH90) and 95 per cent (MH95) of mature height. Other measurements such as sitting height may be similarly treated if desired. The interrelationship between those and other more traditional maturity indices have then been studied in further depth.

For the present study we have used the growth data of children from both the Harpenden Growth Study (Tanner, 1962) and the London cohort of the coordinated growth studies of the International Children's Centre (Falkner, 1980). As one aim was the comparison of bone-age and chronological-age related variables we were restricted to those children where sufficient bone ages had been determined. Further, only those with substantial measurements

were suitable (at least 5 six-monthly measurements before the start of the adolescent growth spurt and 3 similar measurements after the approximate PHV-age); therefore only the data from 39 girls and 51 boys were eventually used. For some measurements the numbers were even fewer due to individual missing observations. Table 1 gives the means, standard deviations and sample sizes for the variables studied.

Before proceeding it was important to compare these mathematically-derived variables with those obtained by graphical means. At an earlier time, Mr. R. H. Whitehouse had prepared graphical estimates of PHV-age in 22 of the overall data set of 90 children. The correlation between the age of PHV by numerical and graphical methods was 0.97 ($P < 0.001$) when boys and girls were considered together.

COMPARISON OF CHRONOLOGICAL AGE AND BONE AGE OF ATTAINING VARIOUS STAGES OF PUBERTY

It is of little interest to show high correlations between different maturational ages of attaining particular events during maturation. For example, it is of little value to show a high correlation between CA and BA of attaining B2. These events are always fairly highly correlated because the principal process that is occurring is the passage of time, and this applies to both maturity measures. To answer the question whether any particular maturity index has a meaningfully closer relationship with the attainment of a particular maturational event, it is more useful to proceed as follows. If the hypothesis was, for example, that the attainment of PHV was more closely associated with a particular BA than CA then the variance for BA of attainment of this event should be less than that for CA. This can be tested by comparing the ratio of the two relevant variances by an F-test. If the ratio is constructed with the CA variance as the numerator and the BA variance as the denominator, then an F statistic greater than 1.0 implies reduction of variance by using BA and implies greater association between the occurrence of the event and the attainment of a particular BA. This was extensively studied by Marshall (1974) who showed remarkably little difference in the association between pubertal events and BA or CA. The major exceptions were the age of attaining menarche, MH95 and PH3 in girls, and MH95 alone in boys.

We have extended this study following very similar methodology using ages for peak height velocity, peak sitting height velocity (PSHV-age) and peak sub-ischial leg length velocity (PSSLV-age), together with ages of attainment

Table 2. The results of variance ratio (F) tests comparing the variance of 'age' of attainment of various pubertal stages when based upon chronological age (CA) or bone age (BA).

		BOYS		GIRLS	
		F	n	F	n
G/B Stage	2	0.78	47	1.17	33
	3	0.79	42	2.29*	30
	4	0.73	35	3.74**	21
	5	1.07	31	1.93	25
PH Stage	2	2.27**	43	1.19	26
	3	1.01	41	2.00*	32
	4	1.48	30	2.30*	26
	5	1.55	33	0.85	21
Menarche				12.01***	39

The ratio was set-up with the variance for CA as numerator so that an F value >1.0 implies a reduction in variance when using BA. The symbols *, ** and *** indicate statistical significance at $P < 0.05$, < 0.01 and < 0.001 respectively.

of 90 and 95 per cent of mature height, all derived from the curve-fitting methods.

Table 2 shows the F-statistics comparing the variance of CA and BA at the attainment of various puberty stages in boys and girls. In this and subsequent tables the sample sizes (n) of some variables may differ slightly from those in Table 1. This is because when calculating the variance ratio it was essential to retain the same children in both numerator and denominator. If there was a missing observation in one then the equivalent result for the same child in the other part of the ratio was deleted. In boys there was no difference in variance between CA and BA, but in the girls there was significantly lower variance in BA of attaining B3 and B4. Indeed, in the girls there was a general trend to lower variance in BA for all stages although not statistically significant for B2 and B5. The pattern is rather similar for PH stages though less marked.

For menarche there was a highly significant reduction in variance for BA such that menarche was achieved at 13.23 ± 0.30 (mean ± SD) years of BA compared with 12.98 ± 1.04 years of CA($P < 0.001$).

Table 3. The results of variance ratio tests comparing the variance of 'age' of attainment of PHV-age, PSHV-age and PSLLV-age based upon CA or BA. Conventions as for Tables 1 and 2.

	BOYS		GIRLS	
	F	n	F	n
PHV-age	1.40	43	1.54	35
PSHV-age	2.35**	32	1.27	33
PSLLV-age	1.23	38	0.89	36

Table 3 shows similar results for the three peak velocity ages. The standard deviation for attainment of peak height velocity based on CA was 0.82 and 0.85 years for boys and girls respectively and when based on BA was 0.69 and 0.68 years. In neither sex was the variance ratio between chronological and bone age basis significant. When PSHV-age and PSLLV-age were similarly studied the remaining results in Table 3 were obtained. Here there was a significant variance ratio of 2.35 ($P < 0.01$) for PSHV-age in boys suggesting that the attainment of this event in boys is more closely related to BA than CA. The other comparisons were insignificant.

In Table 4 we show similar F statistics for comparisons of CA and BA for reaching MH90 and MH95. In boys there seems to be a general reduction in variance when BA is used as the maturity basis; this is most striking for height with $F = 5.86$ ($P < 0.001$). In girls there is no such phenomenon.

In the above series of comparisons we have only studied the PHV variables and MH90 and MH95 as if they were events of puberty comparable to the more conventional puberty stages. It is much more relevant to reverse the situation and consider these size-related events as a basis for a maturity scale. This would naturally take the form of the time elapsed before and after the event. The time could be in units of CA or, in principle, in units of BA; only the former will be considered further here. The attainment of other life events can then be assessed relative to this new scale. To test whether these events are more

closely related to the new scale or to CA or BA, variance ratio tests analogous to those described above can then be performed.

Table 4. The results of variance ratio tests comparing the variance of 'age' of attainment of MH90 and MH95 for height and sitting height based upon CA or BA. Conventions as for Tables 1 and 2.

		BOYS		GIRLS	
		F	n	F	n
Height	MH90	1.30	31	0.59	35
	MH95	5.86***	31	0.83	36
Sitting height	MH90	2.14*	33	1.02	37
	MH95	3.25**	30	1.66	27

Table 5. The results of variance ratio tests comparing the variance of 'age' of attainment of various puberty stages when based upon CA or PHV-age.

		BOYS		GIRLS	
		F	n	F	n
G/B Stage	2	1.16	50	1.06	34
	3	2.53***	50	4.33***	36
	4	1.66*	48	2.06*	26
	5	1.40	45	1.25	37
PH Stage	2	1.63	47	1.54	29
	3	2.88***	47	2.31**	37
	4	2.50**	45	2.39**	35
	5	1.80*	47	1.23	34
Menarche				3.24***	39

The ratio was set-up with the variance for CA as numerator so that an F value >1.0 implies a reduction in variance when using PHV-age. Conventions for statistical significance as in Table 2.

In Table 5 are given the results of the variance ratio tests comparing CA and PHV-age as maturity scales and relating puberty stages to them. The pattern is a little different to that seen when CA and BA were compared. In boys there was substantial reduction in variance for both G and PH stages especially in stages 3 and 4, unlike the pattern with BA where there was no variance reduction. For the girls the general pattern was much closer between Tables 2 and 5 with a general reduction in variance by both BA and PHV-age. A striking difference is that the reduction in variance in 'age' at menarche is much less with PHV-age than BA. Table 6 explores this a little further by comparing the variance of attaining the various puberty stages in terms of BA and PHV-age. In the boys there is clearly a general variance reduction when PHV-age is used as the time scale, which is most striking in G3 and PH3. In girls there is really nothing to choose between the two scales. The same pattern is seen when PSHV-age or PSLLV-age is substituted for PHV-age (not shown).

Table 6. The results of variance ratio tests comparing the variance of 'age' of attainment of various pubertal stages when based upon BA or PHV-age. Conventions as for Table 5.

		BOYS		GIRLS	
		F	n	F	n
G/B Stage	2	1.32	47	0.84	33
	3	3.05***	42	1.69	30
	4	2.21*	35	0.99	21
	5	1.47	31	0.72	25
PH Stage	2	0.71	43	1.50	26
	3	3.08***	41	0.92	32
	4	1.83*	30	0.74	26
	5	1.16	33	1.03	21
Menarche				0.27	35

Tables 7 and 8 carry out the comparable analysis using MH95 in place of PHV-age. In girls there is slight, but consistent variance reduction with MH95 compared to CA, whereas in boys there is none. Comparison of BA and MH95

(Table 8) shows that in boys the variance is lower with MH95 but that there is no difference for girls. Substituting MH90 for MH95 produced a similar but weaker pattern.

Table 7. The results of variance ratio tests comparing the variance of 'age' of attainment of various pubertal stages when based upon CA or MH95. Conventions as for Table 5.

		BOYS		GIRLS	
		F	n	F	n
G/B Stage	2	1.06	50	1.65	34
	3	1.63	50	2.95***	36
	4	1.15	48	2.02	26
	5	0.81	45	1.35	37
PH Stage	2	1.27	47	1.34	29
	3	1.22	47	2.13*	37
	4	0.96	45	2.36*	35
	5	1.01	47	1.18	34
Menarche				4.09	39

One particular feature requires emphasis: when 'age' of menarche is studied, by far the lowest variance is that associated with BA. In Table 2 the ratio between the variance in CA and that in BA is 10.5; in Tables 5 and 7 the comparable ratio is 3.24 and 4.09 respectively suggesting substantially less variance reduction in the cases of PHV-age and MH95. This is confirmed in Tables 6 and 8 when BA and the other scales are compared directly. In the case of PHV-age the F of 0.27 shows that the variance of PHV-age of attaining menarche is nearly four times that of BA; for MH95 it is about three-fold.

CONCLUSIONS

Comparing the results in Table 1 with those of Marshall (1974) shows a similar pattern, but with some slight differences. Marshall's data were taken from the Harpenden Growth Study alone and were not constrained by the

Table 8. The results of variance ratio tests comparing the variance of 'age' of attainment of various pubertal stages when based upon BA or MH95. Conventions as for Table 5.

		BOYS		GIRLS	
		F	n	F	n
G/B Stage	2	1.23	47	1.50	33
	3	2.65***	42	1.16	30
	4	1.90*	35	1.24	21
	5	1.26	31	0.74	25
PH Stage	2	0.55	43	1.32	26
	3	2.47**	41	0.87	32
	4	1.42	30	0.88	26
	5	1.09	33	1.00	21
Menarche				0.36	35

considerations of curve-fitting procedures, which probably explains most of the differences. In the present study the CA of attainment of most pubertal events was a little earlier, especially for the girls. The children from the International Children's Centre study, which were included in this analysis, were generally more advanced than those from Harpenden and this is sufficient to explain the small discrepancy.

Further comparison between the two studies shows an essentially similar pattern when CA and BA are compared. In both cases in boys there was no reduction in variance in BA of attaining G or PH stages compared to CA although the size-related variables, PSHV-age and MH95, were attained with reduced variance in BA. In girls there was generally closer association of the B and PH stages with BA but no reduction in variance of 'age' of attaining the size-related events.

When the maturity scales based on PHV-age and MH95 were studied, a clear pattern emerged. In girls the association between B and PH events and BA, PHV-age or MH95 were about the same. However, menarche was quite different and clearly much more closely related to BA than any other maturity scale.

In boys, PHV-age was most clearly related to G and PH stages and resulted in more variance reduction than did BA or MH95. Overall, it seems that in girls the B and PH events and menarche are most closely related to skeletal maturation and are less influenced by the more directly size-related scales. In contrast, the G and PH stages in boys are best predicted by progression of growth, particularly scales based on PHV-age. Scales derived from 'age' attainment of fixed percentages of mature height seem to be the least related to the evolution of the other pubertal events in either sex.

There is one important point. It is quite clear that there are several ways of assessing maturation and hence tempo. Some are essentially immediate such as BA, and some inevitably retrospective such as PHV-age. However, they do not all measure the same thing and there can be some situations when one scale is more useful and others when another is more appropriate. What remains uncertain is whether there are a number of relatively independant maturational processes, reflected by the scales described above, which combine to govern the tempo of growth of an individual or whether the different scales all reflect a single process measured indirectly with varying degrees of accuracy.

REFERENCES

Demirjian, A. (1978). Dentition, In: F. Falkner & J.M. Tanner (editors), Human Growth, vol.2. New York/London: Plenum Press.

Falkner, F. (ed.) (1980). Growth and Development of the Child, Courrier (International Children's Centre), vol. XXX, Special edition, Paris.

Greulich, W.W. & Pyle, S.I. (1959). Radiographic Atlas of Skeletal Development of the Hand and Wrist, 2nd edition. Stanford: Stanford University Press.

Marshall, W.A. (1974). Inter-relationships of skeletal maturation, sexual development and somatic growth in man. Annals of Human Biology, **1**, 29-40.

Marubini, E. (1984). Review of models suitable for the analysis of longitudinal data, In: J. Borms, R. Hauspie, A. Sand, C. Susanne & M. Hebbelinck (editors), Human Growth and Development. New York/London: Plenum Press.

Preece, M.A. & Baines, M.J. (1978). A new family of mathematical models describing the human growth curve. Annals of Human Biology, **5**, 1-24.

Preece, M.A. & Heinrich, I. (1981). Mathematical modelling of individual growth curves in the study of the control of growth. British Medical Bulletin, **37**, 247-252.

Pyle, S.I. & Hoerr, N.L. (1955). Radiographic Atlas of Skeletal Development of the Knee. Springfield: Thomas.

Roche, A.F., Wainer, H. & Thissen, D. (1975). Skeletal Maturity, The Knee as a Biological Indicator. New York/London: Plenum Medical.

Tanner, J.M. (1962). Growth at Adolescence, 2nd edition. Oxford: Blackwell; Springfield: Thomas.

Tanner, J.M. (1981). A History of the Study of Human Growth. Cambridge: Cambridge University Press.

Tanner, J.M., Whitehouse, R.H., Cameron, N., Marshall, W.A., Healy, M.J.R. & Goldstein, H. (1983). Assessment of Skeletal Maturity and Prediction of Adult Height (TW2 Method), 2nd edition. London: Academic Press.

Todd, T.W. (1937). Atlas of Skeletal Maturation (Hand). St. Louis: Mosby.

Wilkins, L. (1950). The Diagnosis and Treatment of Endocrine Disorders in Childhood and Adolescence. Springfield: Thomas.

BIOLOGICAL AGE ASSESSMENT IN ADULTHOOD

G. A. BORKAN

Normative Aging Study,
Veterans Administration Outpatient Clinic
Boston, Massachusetts, United States of America

INTRODUCTION

Human biologists have devoted considerable effort to devising biological age measurement techniques for childhood, and these have been used with success in evaluating variation in developmental rates. Although the concept of biological age is equally appropriate to the ageing process, there has been much less progress in measurement of this latter phase of the life cycle. The lack of progress is in part due to the newness of gerontology as a scientific discipline, but also to the unique features of the ageing process itself. While the sexual, dental and skeletal development of the child progresses through a series of relatively invariant stages, ageing changes (with the exception of menopause in the female) are progressive rather than stepwise. Further, while most children are healthy, as adulthood progresses the divergence in health becomes considerable, and identifying completely healthy aged individuals is difficult. When the long duration of human ageing is recognised, and is added to the uncertainty about when it begins and ends, it becomes clearer why biological age assessment has been less successful for adults than for children.

Gerontologists have recognized the importance of measuring biological age in adulthood and a number of authors have discussed possible approaches to the problem (Benjamin, 1947; Comfort, 1969; Bourlière, 1970; Shock, 1977). Several investigators have developed biological age scales using diversified batteries of physiological tests. Typically, test parameters are selected based on their cross-sectional linear correlation with chronological age. This approach assumes that those traits which vary most closely with age are the best indicators of the ageing process. In all these studies, the goal has been to compare an individual to his chronological age peers to determine his relative status in selected aspects of ageing. Such procedures essentially attempt to quantify subjective observations, such as a doctor telling a man of 70 that he has the heart of a 40-year-old. The use of quantitative traits as markers, rather

than the discrete stages used in child development assessment, reflects the lack of stepwise marker traits known for ageing. The ability to extrapolate such quantitative assessments to imply actual rates of change depends on the extent to which an individual's status on a given variable indicates change he has undergone, rather than original endowment, random fluctuation, or measurement error.

The approach most used has been to combine a wide range of age-correlated physiological parameters in a multiple regression equation with chronological age as the dependent variable (Hollingsworth et al, 1965; Dirken, 1972; Heikkinen et al, 1974; Furukawa et al, 1975; Webster & Logie, 1976). In these studies the residuals from the regression equation (observed minus predicted scores) are taken as an indication of an individual's biological age. In other words, the extent to which an individual's true age is mispredicted based on his physiological status is his biological age. This methodology, which derives a single overall score, can also be used to develop specialized biological ages such as the determination of anthropometric age (Damon, 1972), auditory age (Bell, 1972), and four other biological ages as derived with data from the V.A. Normative Aging Study in 1972. Factor analysis has been used to truncate a large set of physiological parameters to a set of independent ageing criteria (Clark, 1960; Jalavisto & Makkonen, 1963). However, if the parameters placed into the factor analysis are selected for their diversity, then success in reducing the dimensions of the data will be minimal (Borkan, 1978).

Despite the considerable efforts to develop biological age prediction equations, there has been little attempt to use the derived biological age scores to evaluate the sources of variation in human ageing. Two studies have compared healthy and ill individuals, and found the unhealthy group to be biologically older (Furukawa et al, 1975; Webster & Logie, 1976). Costa and McCrae (1980) used biological age scores to predict subsequent longitudinal change in Normative Aging Study participants, but had little success. Hansen (1973) demonstrated that biological age was a better correlate of periodontal disease than chronological age.

Criticism of the multiple regression approach to biological age in adulthood has come from two different perspectives. Some investigators have maintained that physiological and health parameters, which are most predictive of impending death, rather than predictive of chronological age, should be the ones used to measure biological age. This viewpoint is espoused by Brown and Forbes

(1974a,b) who developed a mathematical model relating progressive physical decline with age to increased likelihood of death. More recently, Brown and Forbes (1976) held that variables selected for inclusion in a biological age test battery must separate the population into subgroups of greater and lesser risk of mortality. This approach, while initially appealing, does not adequately differentiate ageing from disease. It does not differentiate, for example, a 30-year-old cancer victim from a 70-year-old one, each with a one year life expectancy. Further, it only examines those aspects of function related to survival, although many changes of ageing may not be deleterious. This approach may prove more valuable to actuarial and demographic investigations than to evaluation of ageing as a developmental process.

More direct criticism of the multiple regression model comes from Costa and McCrae (1980) who indicated that in previous studies (1) the researchers assume that a significant correlation with age indicates that the parameter changes with age, and does so in a linear manner; (2) the regression approach weights the parameters to give the best prediction of chronological age, which is the very scale which the researchers wish to supercede; (3) the size of the residuals of the regression equation (and thus the biological age scores) is determined by the level of correlation between the independent variables and age; (4) because of the way the regression line is calculated, younger individuals tend to be predicted as older than their age and older individuals are predicted as younger, thus biasing further analyses; (5) the value of the resultant scores as a measure of ageing is the extent to which the error in predicted age is a reflection of actual change undergone by the individual, and not original endowment; (6) the regression approach assumes ageing is a unitary phenomenon, and evidence supports multiple causes and rates of change.

Based on the experience of previous researchers and the criticisms made by Costa and McCrae (1980), we have developed a new approach to assessment of biological age in adults. This method uses a profile of 24 biological age scores reflecting various aspects of physical function. The profile approach allows the possibility that different body systems may age at different rates within the same individual. The purpose of this paper is to describe the use of the biological age profile, demonstrate differences in biological age in men of different marital status, education and self-reported health, and show similarities between biological ages of related individuals.

METHODS

This investigation is based on data collected by the Baltimore Longitudinal Study of Aging of the Gerontology Research Center, National Institute on Aging. This longitudinal study was begun in 1958 and since that time the male participants have returned at 18 month intervals for a comprehensive series of physical evaluations, personal questionnaires, and interviews. Participants are from the Baltimore-Washington area and most are white, college-educated, and middle-income. At entry they ranged in age from 17 to 98 years with an average age of 50. Participants were self-recruited into the study, and no applicant was excluded for health reasons. Further demographic description of this population may be found in Stone and Norris (1966).

For the present study, cross-sectional data for 1086 individuals were analysed. Data used were the first data points available for each variable for each subject, which generally corresponded to the first study visit. The rationale behind the biological age profile and the calculation of biological age scores has been described in detail by Borkan (1978) and Borkan and Norris (1980). A total of 24 physiological parameters were selected as a representative subset of the very large number of characteristic changes of ageing. The criteria for selection were that the variables have a positive or negative linear trend with age in cross-sectional and longitudinal data. Selected parameters had been used in other studies of biological age, had good replicability, and reflected a wide range of physical functions.

The biological age concept implies that for an age-related parameter (with a positive slope, for example), individuals whose scores are above the mean for their chronological age peers are biologically older than those below the mean. That is, their scores are more like those of individuals chronologically older than themselves. This suggests that a biological age score should simply be a statistic which describes an individual's status relative to his chronological age group for each parameter, expressed in standardised units. In statistical terms, the raw data should be corrected for the effects of chronological age, and converted to z-score (standard deviation) units.

The steps in transforming the raw data into biological age scores were:

1. simple piecewise linear regression (Neter & Wasserman, 1974) of each variable on age;
2. subtraction of the predicted score from the actual score of each individual (i.e. calculate residuals);
3. standardization of residual scores using the z-transformation,

accounting for increased variance of data with age;

4. conversion of data so that positive scores refer to greater biological age. Positive sloped variables already were in this form; negative sloped variables were multiplied by -1 to facilitate interpretation.

As a result of these procedures, the raw data for 24 age-related variables were transformed into 24 biological age scores reflecting a man's physical status relative to his chronological age peers. Because the biological age scores are in standardized units, all 24 variables can be compared simply. A further advantage is that men of different chronological ages can have the same biological scores (e.g. ±1 S.D.) for a given variable, indicating that their status relative to their own chronological age peers is identical. Because of this correction, men of different chronological ages can be included in the same analysis without confounding the results.

Biological age scores can be plotted on a profile chart using the technique of Garn (1977). The 24 scores of an individual may be graphed, or subgroups may be plotted by their mean scores for the 24 variables and the pairs of means may be analysed by t-test. Breakdowns of the population based on education, marital status and self-reported health are examined for differences in biological age status in the present report.

RESULTS AND DISCUSSION

The 24 variables selected for inclusion in the biological age profile are listed in Table 1 with their sample size and correlation with age. The quantity of missing data differs among the variables because each participant's visit to the Baltimore Longitudinal Study included only a fraction of the total examination. Certain tests were begun a number of years into the study, and much of the missing data resulted from individuals who did not continue after the initial visit, and therefore did not have all the tests. All the correlation coefficients are statistically significant. Further screening analyses of these parameters (Borkan, 1978) indicated that longitudinal slopes paralleled cross-sectional trends for these variables, demonstrating that secular or cohort trends did not account for the apparent age-relatedness of these variables.

The subject population was divided into two groups for comparison based on education: those who had a junior college diploma or less (264 individuals), and those with at least a bachelor's degree (787 individuals). It was hypothesised that the less-educated group would be biologically older than the more educated group, because more educated people have better health care, longer life

Table 1. Sample size and correlation coefficients with age for variables selected for biological age test battery

Variable	N	r
Forced expiratory volume (1 sec)	969	-.698**
Vital capacity	971	-.606**
Maximum breathing capacity	1029	-.547**
Systolic blood pressure	1077	.538**
Diastolic blood pressure	1077	.368**
Haemoglobin	1071	-.223**
Serum albumin	777	-.356**
Serum globulin	777	.092*
Creatinine clearance	1051	-.602**
Plasma glucose (OGT test)	739	.279**
Auditory threshold (4000 cps)	862	.549**
Visual acuity	940	-.306**
Visual depth perception	935	.232**
Basal metabolic rate	1035	-.337**
Cortical bone %	983	-.435**
Creatinine excretion	1080	-.538**
Hand grip strength	943	-.501**
Maximum work rate	892	-.511**
Benton visual memory test (errors)	905	.502**
Tapping time (medium targets)	992	.468**
Tapping time (close targets)	991	.366**
Reaction time (simple)	687	.287**
Reaction time (choice)	701	.220**
Foot reaction time	734	.222**

*p <.05; **p <.01

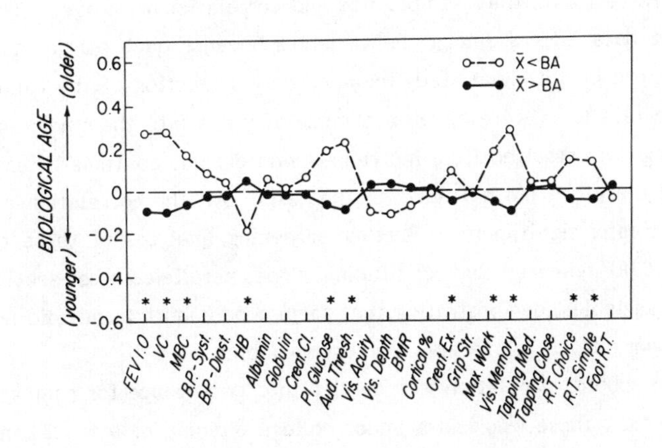

Figure 1: Biological age profile based on educational status (>BA = bachelor's degree or higher). Mean scores for subpopulations compared by Student's t-test, * signifies p <.05.

expectancy, and safer work conditions. However, it is possible that less-educated individuals could be at an advantage if they are more physically active on the job. The results of this analysis (Fig. 1) show that the less-educated group was biologically older in 17 of 24 variables, and 10 of these comparisons are statistically significant. Significant differences occur for the three lung measures, plasma glucose, auditory function, creatinine excretion, maximum work rate, visual memory, and both auditory reaction times. These comparisons indicate that the less-educated group was biologically older in parameters that reflect physiological function and health (the left side of the profile), were capable of less work, and slower on psychomotor tests. Their poorer visual memory may reflect the influence of education on scores on this test.

Reversals of expectation, i.e. the finding that the less-educated group was biologically younger, were significant for haemoglobin and non-significant for visual measures. It is interesting that less-educated individuals were biologically younger in vision, but biologically older in auditory measures. This may reflect hazards of their work environment. In summary, the biological age profile of less-educated individuals indicates generally greater biological age on physiological variables and on psychomotor tasks. Their body composition was very slightly more youthful, meaning that they were somewhat larger in body size than the more educated group.

In terms of marital status, the study population again is atypical of the average U.S. population, in having greater marital stability. At the time of examination 85% were married, 5% had never married, and 10% were widowed, separated or divorced. The two subpopulations used for comparison were the group of men who were presently married and the group which had never married or were widowed, separated or divorced. These two groups contain all the individuals for whom marital information was available.

We hypothesised that non-married individuals would have generally greater biological age because married men tend to have greater personal stability and longer life expectancy than non-married men. We found that on 19 of 24 profile variables, the results were in the expected direction, with non-married men having greater biological age than married men (Fig. 2). However, cortical bone percentage was the only significant difference, although moderate differences were observed for lung function, blood pressure and psychomotor tests. Of the five variables which reversed expectation none approached statistical significance.

From this analysis it is evident that non-married men tended to be

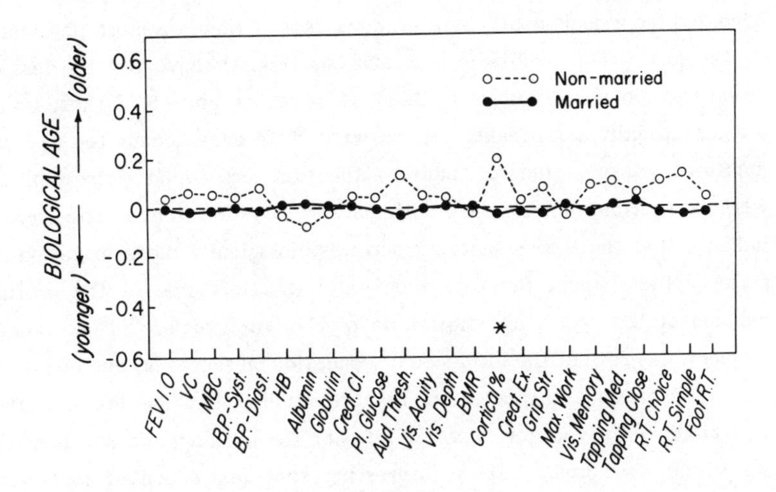

Figure 2: Biological age profile based on marital status. Mean scores for subpopulations compared by Student's t-test, * signifies $p < .05$.

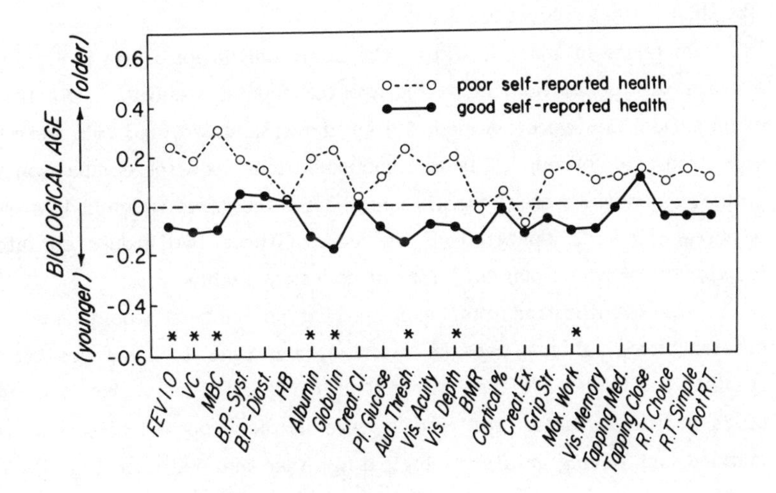

Figure 3: Biological age profile based on self-reported health. Mean scores for subpopulations are compared by Student's t-test, *signifies $p < .05$.

biologically older than married men on many ageing parameters. While this may be due to the beneficial influence of married life, it is also possible that individuals who never marry or who do not remain married are a subset of the total population with less favorable health characteristics to begin with.

One of the questionnaires filled out by subjects in the Baltimore Longitudinal Study was "Your activities and attitudes" by Burgess, Cavan and Havighurst. As directed in the scoring manual, the scores on four of the questions were combined to yield a rating of general health and vitality. The subpopulations for comparison were the 15% of the sample with the poorest self-reported health (3.5 points or less) and the 15% with the best self-reported health (9.5 points or more).

We hypothesised that individuals whose health was poor might be biologically older than those in good health, because it is a frequent observation that illness accelerates the ageing process, at least on the basis of visual estimation. This hypothesis was vigorously suported by the profile comparisons (Figure 3) in which all 24 variables were in the expected direction. Significant differences were found in the three lung variables, two serum proteins, hearing, vision, and maximum work rate.

The results of this comparison demonstrate that the mean profile of individuals in poor health was biologically older than for those who were healthy. Since illness and ageing are positively correlated, the close relationship between biological age and health is not surprising. Whether illness actually accelerates the process of ageing or merely simulates its appearance is not demonstrated in the pattern profile analysis. What is shown is that unhealthy people have the physical, sensory, and psychomotor traits of persons chronologically older than themselves.

The foregoing analyses demonstrate the association of biological age with personal background and lifestyle. We also examined the heritability of biological age using the biological age profile technique. To study this problem, 96 father-son and 51 brother-brother pairs in the population were identified. This large number of related individuals may be traced to the mode of subject recruitment which allowed participants to recommend the study to friends and relatives.

We correlated the scores of fathers and sons for each of the 24 profile variables (cross-sectional data) and did the same for brother-brother pairs. The correlations and their significance levels are given in Table 2. We expected that the correlations between pairs of unrelated individuals would be zero, and that

Table 2. Correlation of biological age scores between related individuals

Variable	Father-Son N	R	Sig.	Brother-Brother N	R	Sig.
Forced expiratory volume	91	.144	*	48	.006	
Vital capacity	91	.248		48	.151	
Max. breathing capacity	94	.061		50	.196	
Systolic blood pressure	95	.126		50	.277	
Diastolic blood pressure	95	.079		50	.209	
Haemoglobin	94	.129		50	.365	**
Albumin	62	.305	**	37	.443	**
Globulin	62	.300	*	37	.271	
Creatinine clearance	93	.100		51	.037	
Plasma glucose	63	.120		34	.134	
Auditory threshold	79	.311	**	45	.069	
Visual acuity	91	.163		47	.096	
Depth perception	91	.031		47	.092	
Basal metabolic rate	92	.039		48	.306	*
Cortical bone %	92	.319	**	43	.445	**
Creatinine excretion	95	.308	**	51	.211	
Grip strength	81	.071		37	.088	
Maximum work rate	78	.005		36	.158	
Visual memory	85	.115		40	.237	
Tapping rate (MN)	91	.430	**	49	.490	**
Tapping rate (SN)	91	.363	**	49	.456	**
Reaction time (choice)	56	.337	*	35	.182	
Reaction time (simple)	56	.095		36	.185	
Foot reaction time	63	.318	*	40	.131	

* p <.05; ** p <.01

significant positive correlations would suggest a familial and possibly genetic aspect to biological age.

All but one of the correlations in Table 2 were positive and a substantial number reached statistical significance. High correlations were found for several of the blood biochemistry measures, for several body composition measures, and for tapping and reaction time tests. The differences in correlation levels between variables suggest that the familial component of some aspects of ageing may be greater than for others. It should be remembered that these results may reflect either common genetic endowment, or common home environment and shared lifestyle. Therefore, such correlations do not prove that genetic inheritance is a factor in biological age status, though it is clearly a strong possibility.

The foregoing comparisons demonstrate statistically significant differences in biological age status between individuals. However, it is

important to bear in mind the assumptions and limitations of the approach. The technique is based on the view that an individual's scores on age-related variables, when compared to his chronological age peers, are an indication of his biological age. This is very similar to when a physician tells a patient he has the lung capacity of someone 20 years younger. In this case the physician is making the suggestion that the patient has changed (aged) less than others of his same age group. Of course, the patient may have always had a high score on this parameter due to genetic endowment, large body size, athletic training, or pure chance. The extent to which the man's lung function reflects his actual rate of pulmonary change is the extent to which such an observation is a good measure of ageing.

Another problem, true of all ageing research, is the differentiation of ageing from disease. It is well accepted that advancing age is associated with increased incidence of many diseases. There are few parameters in the biological age profile which could not be viewed as aspects of health as well as ageing, and none of which truly poor scores would not be of medical concern. Considerable debate could be undertaken to prove whether this or any ageing profile measures ageing or disease. But from a practical standpoint such an argument is moot, because these are all measures of functional status in late adulthood, and are indications of relative viability of the individual. And ultimately, the purpose of ageing research is to improve health and well-being in the last part of the life cycle. Whether these variables are measures of ageing or disease therefore becomes less important.

The approaches reported here are one way in which it is possible to examine factors which may possibly influence the course of ageing. The biological age test battery proposed is not immutable, but can be composed of any measures that are indicative of ageing. Indeed, there could be an entire profile devoted to visual function or exercise performance, rather than the broad profile described. The important consideration is that the parameters in the profile comprise significant aspects of the ageing process. Particularly important would be inclusion of parameters which represent the basic biochemical "causes" of ageing. When these underlying processes of ageing are identified, they may be incorporated into a profile technique such as the one described here, to test pharmacological agents or behavioural traits which may influence the course of the ageing process.

REFERENCES

Bell, B. (1972). Significance of functional age for interdisciplinary and longitudinal research in aging. Aging and Human Development, **3**, 145-148.

Benjamin, H. (1947). Biologic versus chronologic age. Journal of Gerontology, **2**, 217-227.

Borkan, G.A. (1978). The assessment of biological age during adulthood. (Doctoral dissertation, University of Michigan.) Dissertation Abstracts International, 39/06-A:3682 (University Microfilms No. 78-2286).

Borkan, G.A. & Norris, A.H. (1980). Assessment of biological age using a profile of physical parameters. Journal of Gerontology, **35**, 177-184.

Bourlière, F. (1970). The assessment of biological age in man. Geneva: World Health Organisation Paper No. 37.

Brown, K.S. & Forbes, W.F. (1974a). A mathematical model of aging processes, I. Journal of Gerontology, **29**, 46-51.

Brown, K.S. & Forbes, W.F. (1974b). A mathematical model of aging processes, II. Journal of Gerontology, **29**, 401-409.

Brown, K.S. & Forbes, W.F. (1976). Concerning the estimation of biological age. Gerontology, **22**, 428-437.

Clark, J.W. (1960). Aging dimension: a factorial analysis of individual differences with age of psychological and physiological measurements. Journal of Gerontology, **15**, 183-187.

Comfort, A. (1969). Test battery to measure aging rate in man. Lancet, **2**, 1411-1415.

Costa, P.T. Jr. & McCrae, R.R. (1980). Functional age: a conceptual and empirical critique. In: G.S. Haynes & M. Feinleib (eds), Epidemiology of Aging. NIH Publ. No. 80-969.

Damon, A. (1972). Predicting age from body measurements and observations. Aging and Human Development, **3**, 169-173.

Dirken, J.M. (1972). Functional age of industrial workers. Groningen, Netherlands: Wolters-Moordhoff.

Furukawa, T., Inoue, T.M., Kajiya, F., Takeda, H. & Abe, H. (1975). Assessment of biological age by multiple regression analysis. Journal of Gerontology, **30**, 422-434.

Garn, S.M. (1977). Patterning in ontogeny, taxonomy, phylogeny, and dysmorphogenesis. In: R.K. Wetherington (ed.), Colloquia on Anthropology. The Fort Burgwin Research Center, **1**, 83-106.

Hansen, G.C. (1973). An epidemiologic investigation of the effect of biologic age on the breakdown of periodontal tissues. Journal of Periodontology, **44**, 269-276.

Heikkinen, E., Kishkinen, A., Kayhty, B., Rimpela, M. & Vouri, I. (1974). Assessment of biological age: methodological study of two Finnish populations. Gerontologia, **20**, 33-43.

Hollingsworth, J.W., Hashizuma, A. & Jablon, S. (1965). Correlations between tests of aging in Hiroshima subjects: an attempt to define "physiologic age". Yale Journal of Biology and Medicine, **38**, 11-36.

Jalavisto, E. & Makkonen, T. (1963). On the assessment of biological age. I. A factor of physiological measurements in old and young women. Annales Academiae Scientiarum Fennica, **100**, 1-34.

Neter, J. & Wasserman, E. (1974). Applied Linear Statistical Models. Homewood, Illinois: Richard Irwin.

Shock, N.W. (1977). Indices of functional age. Paper presented at conference "Aging: a Challenge for Science and Social Policy", Vichy, France (April 24-30, 1977).

Stone, J.L. & Norris, A.H. (1966). Activities and attitudes of participants in the Baltimore Longitudinal Study. Journal of Gerontology, **21**, 575-580.

Webster, I.W. & Logie, A.R. (1976). A relationship between age and health status in female subjects. Journal of Gerontology, **31**, 546-550.

SKELETAL AGE AND PALAEODEMOGRAPHY

T. I. MOLLESON

Department of Palaeontology,
British Museum (Natural History), London, U.K.

INTRODUCTION

> *"The most reliable way of assessing the age of an individual from an anatomical point of view is by examination of the skeleton and the teeth. No other part of the body, however carefully examined, can give as much information. The younger a person is, the easier it is to assess age accurately from the bones."* (Harrison, 1958.)

It is widely accepted that expectation of life in the human species has increased not only in the last hundred years for which documentation exists but also for the thousands of years before that. It is equally accepted that human specific longevity, that is the maximum potential lifespan, has not increased, merely the number of people that die before reaching seventy has been reduced. Yet when we look at the age structure of cemetery populations of the mediaeval, Romano-British or Neolithic periods there is little sign of this older age group.

It is easy to calculate a constant 3% per annum death rate for a static, non-replacing population, and this shows that as much as 16% of the skeletons in a cemetery could be over 60 years of age. In reality, mortality rates vary with age and always there is an increase in the number of old dying. I think that we have to find this group before we can attempt any palaeodemographic reconstruction from cemetery material. Human mortality patterns are unlikely to have changed so radically. It is more probable that the fault lies with our methods. In this paper I should like to examine some of our methods of ageing skeletal material and discuss some of the premises that lie behind their acceptance.

Ageing phenomena are the product of a number of progressive changes wrought on the individual skeleton; no two individuals are alike in their rate of ageing; biological variability is at least as great in the human species as it is in any other animal species and a given skeletal age, that is, the maturational or degenerative state of the skeleton, may be attained over a broad chronological

time depending on genetic and environmental factors. In the analysis of historical material which invariably comes without birth certificate and only exceptionally with tombstone information we are utterly dependent on studies of known age material to calibrate skeletal age. Such calibrations beg the question that they are valid for past populations and particular care must be taken before we draw conclusions as to age or rate of maturation in earlier peoples.

AGEING PROCESSES

Methods of ageing skeletal material derive from four different categories of processes working on the hard tissues of the body (Table 1). The first two categories - developmental, and growth and enlargement - give criteria by which the age of juveniles can be assessed; the second two - degenerative and wear, and remodelling - provide methods for the ageing of adults. From the applications of these methods we can identify a number of stages which can be broadly correlated with expected chronological age (Table 2). It is the accuracy and precision of these correlations that is constantly under test and adjustment in our laboratories. In particular, skeletons from the Romano-British (1st-5th Century AD) cemetery at Poundbury Camp, Dorchester, Dorset have been studied for developmental and ageing changes. I present some of our preliminary results here. A fifth category of change resulting in variation in trace element content of bones and teeth is not adequately developed and will not be discussed in this paper. Nor shall I discuss other destructive techniques including osteone senescence and cortical resorption. Such methods have yet to be properly tested.

METHODOLOGY

The most reliable methods of ageing the skeleton are based on the development and eruption of the dentition. The synchondrosis of bones that have developed from one or more primary centres and the fusion of epiphyses to long bone diaphyses are also important indicators of skeletal maturity. Stages of maturation of the skeleton or teeth are scored according to the earliest possible age that the stage can be attained. This gives a minimum age. The maximum age is assessed from the latest time the stage is reached. The age is expressed as a bracket of the minimum and maximum age.

Fazekas and Kosa (1978) have put forward criteria by which the potential viability and maturity of a fetus can be ascertained. It would appear that viability is more closely related to bone development than to size. Neonates from archaeological finds, although smaller than their modern counterparts, can

Table 1. Ageing processes as a basis for assessing skeletal age

Developmental	*References*
Dental development	
Deciduous dentition	Stack & Halazonetis, 1973; Lunt & Law, 1974.
Permanent dentition	Demirjian et al, 1973, 1980; Moorrees et al, 1963.
Ossification and synchondrosis	
Fetus	Fazekas & Kosa, 1978.
Juvenile	Greulich & Pyle, 1959; Tanner et al, 1962, 1983; Garn et al, 1964; Krogman, 1962.
Growth and Enlargement	
Bone growth	
Fetus	Fazekas & Kosa, 1978; Scheuer et al, 1980.
Juvenile	Maresh, 1955; Stloukal & Hanakova, 1978; Sundick, 1978.
Sinus enlargement	Maresh, 1940; Brown et al, 1984.
Degenerative and Wear	
Dental attrition	Zuhrt, 1955; Brothwell, 1963; Gustafson, 1950.
Pubic symphysis wear	Todd, 1921; McKern & Stewart, 1957; Gilbert & McKern, 1973; Nemeskeri et al, 1960.
Progressive osteoarthritis	Stewart, 1958; Sager, 1969; Whittaker et al, 1984b; Stloukal et al, 1975a; Stloukal et al, 1975b.
Remodelling	
Continuous eruption	Levers & Darling, 1983; Whittaker, et al, 1984a.
Cranial suture closure	Manouvrier, 1894; Martin & Saller 1957.
Spongiosa density	Schrantz, 1933; Hansen, 1953; Nemeskeri et al, 1960.
Cortical thickness	Virtama & Helela, 1969; Spencer & Coulombe, 1966.

be assessed as being premature or full-term from the examination of bones of the skull (Table 2 and Plate 1).

The state of fusion of the epiphyses of the long bones is widely used in assessing the age at death of adolescents, since teeth are of little value once the second molar has erupted and before there is significant wear on the teeth; but chronological age can only be inferred from skeletal age in very broad terms because of a number of factors. Principally, individual variation in living

Table 2. Indicators of skeletal age

Fetal and neonate	
(a) Basi-occipital longer in the mid-sagittal plane than wide	Non-viable fetus (less than 28 weeks)
(b) Lesser wing of sphenoid and body fused	Potentially viable
(c) Lesser wing of sphenoid more than twice its width	Potentially viable
(d) Petrous bone fused with squamous part of temporal bone	Mature fetus
(e) Tympanic ring fused to temporal	Mature fetus
(f) Crown of maxillary central incisor (dil) complete	Full term
(g) Occlusal surface of first molar (M1) complete	Full term
(h) Ossification centres of distal epiphysis of femur, proximal epiphysis of tibia, talus, calcaneus and cuboid present	Full term
Infant (less than 12 months)	
(a) Mandible halves not fused	Less than 12 months
(b) Greater wing and body of sphenoid fused	Over 9 months
(c) Crown of lateral incisor (di2) complete	Over 3 months
(d) Crown of first molar (dm1) complete	Over 6 months
(e) Crown of canine (dc) complete	Over 9 months
(f) Crown of second molar (dm2) complete	Over 11 months
Child (pre-puberty)	
(a) Crowns of all deciduous teeth complete	Over 11 months
(b) Mandible halves fused	Over 12 months
(c) Roots of all deciduous teeth complete	Over 3½ years
(d) Frontal halves fused	Over 2 years
(e) Lateral part of occipital fused to basal part	Over 3 years
(f) Odontoid ossicle of axis vertebra fuse	4-6 years or 12 years
(g) Neural arch fused to vertebral body - cervical	Over 3 years
(h) Neural arch fused to vertebral body - lumbar	Over 6 years
(i) Pubis and ischium fused	Over 6 years
(j) Distal end of radius unfused	Pre-puberty
(k) Roots of M2 incomplete	Pre-puberty
Adolescent (puberty)	
(a) Triradiate of acetabulum ossifies	12-16 years
(b) Calcaneus epiphysis fuses	Puberty
(c) Fibula, proximal epiphysis fuses	16 years
(d) Elbow, ankle, proximal femur, metacarpals, metatarsals, phalanges epiphyses fuse	About 18 years
(e) Humerus head, knee epiphyses fuse	About 18 years
(f) Acromion, coracoid process of scapula fuse	Puberty - 20 years
(g) Clavicle, medial epiphysis unfused	Under 25 years
(h) Secondary centres of ribs, vertebral bodies, lumbar vertebrae, sacrum - auricular epiphyses	16-25 years

/contd.

Table 2, contd.

Young adult	
(a) Basi-occipital fuses with basi-sphenoid	18-25 years
(b) Clavicle, medial epiphysis fuses	18-31 years
(c) Iliac crest, ischial tuberosity, inferior iliac spine epiphyses	Appear at puberty, fuse at 25 years
(d) Sacrum, upper bodies fuses	Over 16 years
(e) Sacrum, lower bodies not fused	Under 30 years
(f) Radius distal epiphysis fused	Post-puberty
Mature adult	
(a) Coronal suture closed on endocranial surface	Over 30 years
(b) Sagittal suture closed on endo- and ectocranium	Over 40 years
(c) All cranial sutures closed both endo- and ecto-cranially	Over 55 years
(d) Osteophytosis of vertebrae	Over 30 years
(e) Apex of medullary cavity reaches above epiphyseal line of humerus head	Over 37 years
(f) Apex of medullary cavity extends above lesser trochanter; medial rarefaction of neck, diaphysis border and head of femur	Over 32 years
(g) Symphyseal face of pubis smooth, concave, porous; crest along dorsal and ventral rim	Over 41 years

populations is not insignificant and we have no reason to believe otherwise for past populations. There is a good deal of variation in the standards published by different authors even for the same ethnic group. Racial variation is considerable (Tanner et al, 1983) and we do not always know which ethnic group we are studying. On average there is up to two years difference in development between the sexes. This, together with the fact that a normal individual can deviate from the mean by ± 2 years gives a possible age range for a particular bone stage of no less than six years, since it is rare that we can sex sub-adolescent skeletons (Krogman, 1962).

Garn et al (1964) proposed a system of weighting whereby the appearance of those bones of the wrist and ankle which can be shown to have a high predictive value in age determination are used in preference to, or even to the exclusion of, other bones aberrant or variable in their appearance. They maintain that to consider all bones actually gives less accurate ageing than age determination based on the selected bones that have been shown to be reliable. The epiphyses of the metacarpals and selected digits in the hand, and the metatarsals and selected proximal segments of the digits in the foot provide 19

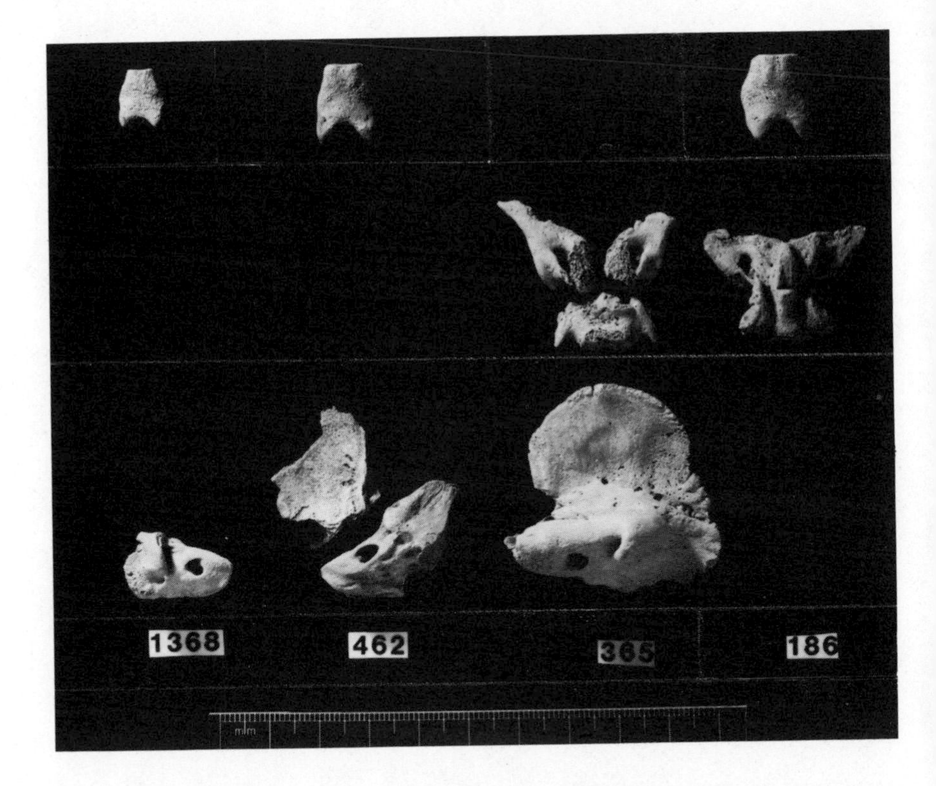

Plate 1. Skull bones of neonates from Poundbury Camp, Romano-British Cemetery. Top row - basi-occipital; middle row - sphenoid; bottom row - temporal bones.

1386: a fetus of less than 28 weeks. It would not have been viable. The basi-occipital is longer than it is wide. The petrous bone is short and rounded.

462: a stillborn fetus of about 36 weeks. The lesser wings of the sphenoid are separate and not fully elongated.

365 and 186: were probably mature fetuses born at term. In 365 the petrous bone was united with the squama of the temporal bone. In 186 the basi-occipital is wider than it is long and the lesser wings of the sphenoid have fused with the body.

centres having a high predictive value. Other bones, such as the triquetral that are extremely variable in time or appearance of the cuboid and the lateral cuneiform that have virtually no predictive value regarding the time of ossification, can yield misleading ages (variant ± 2 years). The method is not, however, very useful for skeletal material since the bones and epiphyses concerned frequently are not recovered.

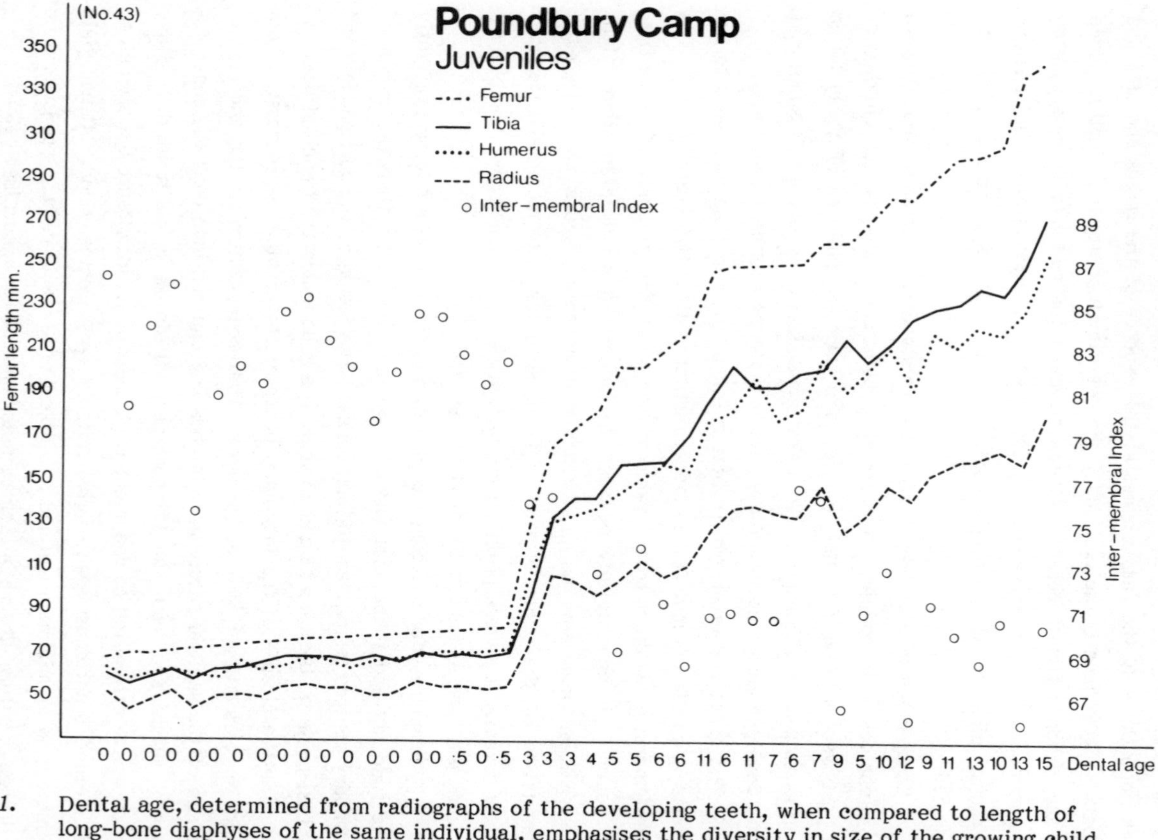

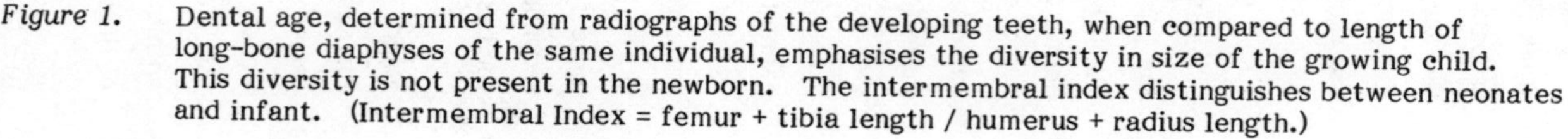
Figure 1. Dental age, determined from radiographs of the developing teeth, when compared to length of long-bone diaphyses of the same individual, emphasises the diversity in size of the growing child. This diversity is not present in the newborn. The intermembral index distinguishes between neonates and infant. (Intermembral Index = femur + tibia length / humerus + radius length.)

Tanner et al (1983) are inclined to take the opposite view to Garn. They have developed a system of weighted scores. Since many bones of the hand and wrist give the same information about maturity, to consider all would give too much importance to the finger bones at the expense of the radius and ulna and carpals. Consequently digits 2 and 4 are not examined at all. Further, since the carpal bones give different information about the maturity process than do the long bones, three separate scoring systems have been derived. One concerns the radius, ulna and finger bones, another the carpals only, and the third both combined.

Comparison of limb length with modern standards such as those of Maresh (1955) enables us to conclude that for all populations studied so far, children of the past were smaller for their (dental) age than are children of today by as much as two years (Figure 2). To make this deduction we have to assume that dental development proceeded in the past at the same rate as it does today.

The greater rate of growth of the femur and the tibia compared to the humerus and radius is manifest in modern children after two years of age. The intermembral index drops from a value of over 80 at birth to below 76 in the first two years. It can usefully be used to indicate a small infant that might otherwise on bone size be mistaken for a neonate (Figure 1). Detailed examination of the bones of small infants with low intermembral index has usually revealed a severe proliferative periostitis. Only one showed no obvious pathology; it just had unusually short forearms.

Synchondrosis of two bones as part of maturation appears to occur slightly more suddenly and to be less variable between the sexes than the fusion of secondary centres (epiphyses) to long bones. The fusion of the basisphenoid and the basi-occipital between 18 and 25 years is a good indicator of adult status.

Suture closure of the individual bones of the cranium is progressive from about 25 years to over 60 years of age. The process starts on the endocranial surface of the sagittal suture near lambda and can continue until the sphenoid-parietal suture and even the temporo-parietal suture is obliterated. Many workers have attempted to document the stages of suture closure but used alone this method is not considered to give reliable ages (Comas, 1960). It can yield an indication of minimum age and Nemeskeri et al (1960) include a consideration of degree of suture closure in their complex method for ageing (see below).

The dental attrition grades of Zuhrt (1955) have been summed by Brothwell (1963) for the three molars to typify stages of wear characteristic of broad age categories. Both Brothwell and Miles who produced his own scheme in

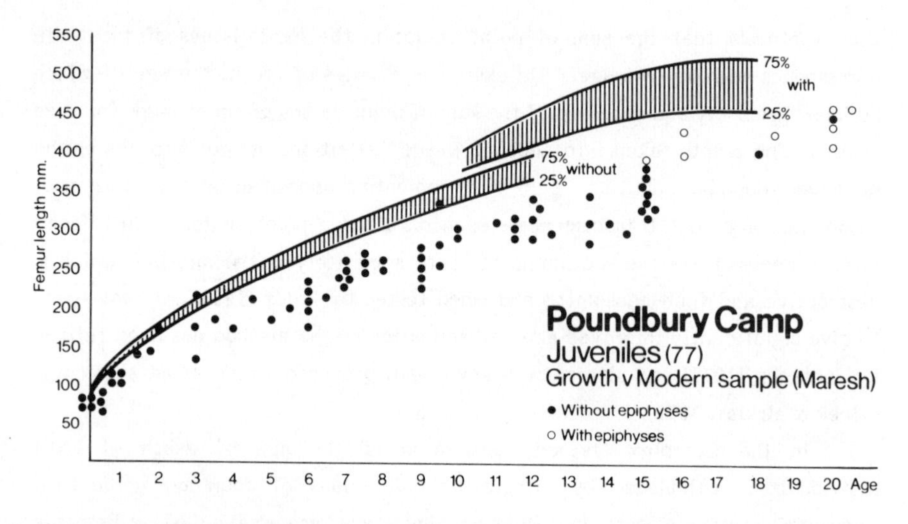

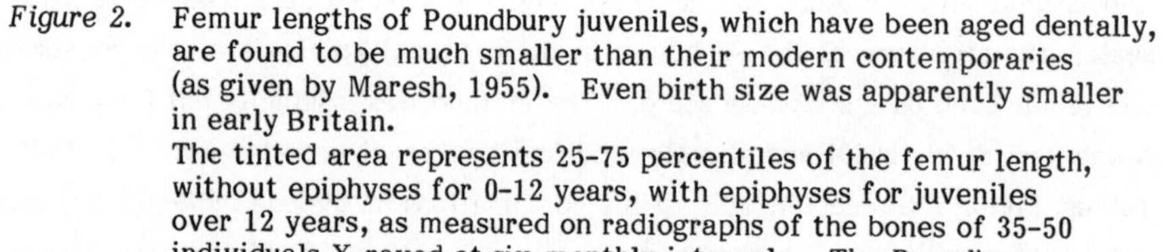
Figure 2. Femur lengths of Poundbury juveniles, which have been aged dentally, are found to be much smaller than their modern contemporaries (as given by Maresh, 1955). Even birth size was apparently smaller in early Britain.
The tinted area represents 25–75 percentiles of the femur length, without epiphyses for 0–12 years, with epiphyses for juveniles over 12 years, as measured on radiographs of the bones of 35–50 individuals X-rayed at six-monthly intervals. The Poundbury bones were measured directly, the difference being about 5% less.

1963 stressed that the rate of occlusal wear in each population should be established before age calculations are attempted. Brothwell's method is probably the predominant method of ageing skeletal material from archaeological contexts used in Britain. It is unfortunate that such a readily applicable method should have been accepted so uncritically by so many. Attrition is caused by wear on the teeth during mastication and the rate is obviously affected by different diets and different methods of preparing food. It is also apparent that the stages of wear at present in use do not represent equal amounts of wear on the teeth. Some stages need to be further divided, others merged to make for a more even scale.

Degenerative changes, tooth attrition, pubic symphysis wear, cortical thinning and trabecular thinning are usually scored according to a system of more or less clearly defined grades. Because different processes proceed at varying rates in different parts of the skeleton, a number of authors have proposed that the grade recorded for each factor should be evaluated on a points system which is then calibrated with an age scale. Gustafson (1950) developed

the hypothesis that the sum of point values in the hard tissues of the tooth corresponds with a given age. He examined a series of teeth, the age of which he knew precisely and determined the sum of point values of six changes for each tooth. The points taken into account were: attrition; gingival to cementum distance; secondary dentine - pulp size; cementum apposition at the root; root resorption; and root transluscence - each on a three point scale. Gustafson's method necessitates the sectioning of teeth accurately in the mid-line which is destructive and time-consuming and when tested by Miles (1963) was not found to give significantly improved age determination. The method has been refined by Johanson (1971) and alternatives have been proposed which avoid sectioning (Vlcek & Mrklas, 1975).

In the Complex Method, estimation of the age of death of adult individuals is calculated by integrating the values of four age indicators. Endocranial suture closure and pubic symphysis wear are assessed on a five phase scale; the structure of the spongiosa in the proximal epiphysis of the humerus and femur each on a six phase scale. The method was originally put forward by Nemeskeri et al (1960) and slightly modified by Acsadi and Nemeskeri in 1970. Tables, taking into account all possible combinations of the age indicator phases, were calculated by Sjovold (1975) so that ages can be read off corresponding to each phase of each age indicator or combination of indicators. The scheme was adopted by the Workshop of European Anthropologists and forms the basis of their Recommendations of Age and Sex Diagnoses of Skeletons (1980). The original reference collection on which the phasing of the age indicators was established consisted of 105 skeletons, mostly accident victims of known age ranging from 21 to 93 years. There were 61 males and 44 females. The sexes are not separated and occupation or dietary status have not been taken into consideration.

Cranial suture obliteration is evaluated according to the methods of Manouvrier (1894) and Martin and Saller (1957). The pubic symphysis is aged according to the method of Todd (1920, 1930) although the original ten phases are redefined in five. The methods of Schranz (1933) and Hansen (1953) are used in the study of the humerus and femur respectively. It is clear from the results that although the phases for each age indicator are progressive, the age of onset and age of progression to the next phase are only in the broadest terms correlated with age. The age changes proceed slowly in the beginning, then the process is accelerated and the phases apparently follow one another at short time intervals. Nemeskeri et al were fully aware of the tentative nature of their

method based as it is on a small sample, particularly in the younger age range. Nor has the method been tested on material from other periods in time or environments. So the tendency to adopt the Complex Method or rather the tables derived from it must be considered premature and other ageing methods must be sought, tested and used.

Krogman (1962) argues that biological age in the skeleton is the sum of *all* skeletal age changes. This attitude is implicit in the approach of such workers as Gustafson, Tanner and Nemeskeri but not of Garn or Fazekas and Kosa. There are some turning points in maturation, such as puberty, and a number of convenient markers: the shape of the basi-occipital of the premature fetus, the synchondrosis of the basi-occipital to the basi-sphenoid in the adult. These markers may also indicate turning points, but skeletal maturation and ageing being continuous processes over such a broad spectrum probably can only be assessed by summing all the skeletal age changes.

Chronological ages which are important socially and demographically may be best indicated by reference to designated markers, puberty for example (Table 3).

Scaling

Skeletal changes, which are progressive in their development, can only be used for the estimation of age after the changes have been graded and the progression of one grade to another scaled. It cannot be assumed that the scale

Table 3. Skeletal indicators of puberty*

Pre-puberty	*Post-puberty*
GIRLS (menarche mean 13.08 years)	
Roots M2 incomplete	Canines, premolars, M2, erupted
Crown M3 incomplete	M3 erupted
Radius distal epiphysis unfused	Radius distal epiphysis fused
Middle third phalange epiphysis unfused	MP3 fused
Distal third phalange epiphysis unfused	DP3 fused
BOYS (male voice mean 15.02 years)	
Not all M2s erupted	Roots of M2 complete
	Roots of M3 half complete
Radius distal epiphysis unfused	Radius distal epiphysis fused
Middle third phalange epiphysis unfused	MP3 fused
Distal third phalange epiphysis unfused	DP3 fused

* from data in V. Hagg & J. Taranger (1982), Amer.J.Orthod.

will be of equally spaced units.

Ideally a regression should be established for each progressive method of ageing and for each population studied. Excessive extrapolation from points should be avoided since the regression is almost certainly not linear for all ages; nor should end points be used to define a regression since these represent extremes of the population. This may seem obvious but at least one ageing study has foundered because the investigator failed to appreciate the dangers of joining in a single regression the values from three motor cycle accident victims in their twenties to the values from 74 geriatric patients. The unevenness of the scale for dental attrition grades of one population is probably the explanation for the poor agreement of age assessed by this method and others such as the state of the pubic symphysis. The later part of the scale is condensed. Those who died in the fourth and fifth decades of life are underscored, the old are underrepresented and the mean age at death is underestimated. That this has far-reaching consequences for palaeodemographic studies must be obvious.

Calibration and resolution

It is one thing to establish a mean value for a given maturity parameter, whether it is dental development, skeletal maturity or degenerative change, and quite another to use these mean values to infer chronological age. The age range over which any one phase may occur may be months, years or decades. It is this that limits the resolution of skeletal ageing. Growth studies on the living show that the size range for juveniles of any given year is greater than the difference between consecutive years and may cover as much as three years. An eleven-year-old may be expected to have a femur length between 332 and 432 mm (diaphysis) but a femoral length of 380 mm could be that of a child aged anywhere between nine-and-a-half and thirteen years. Thus attempts to calibrate size with chronological age cannot hope to achieve great precision. The problems of ageing adults are if anything greater, although accurate ageing probably matters less in most circumstances.

Bone maturation stages and their individual sequences are the same in all populations and are unaffected even by starvation but *rate* of skeletal maturing does reflect the interaction of genetic and environmental forces. Thus populations differ both in mean skeletal maturity at a given age and in the pattern of increments from age to age. Skeletal maturation is slower everywhere in the worse-off compared to the better-off socio-economic groups. A typical difference is from about 0.5 year bone age in early childhood rising to 1.0 'year'

at puberty. However, the amount naturally depends on the breadth of the nutritional and hygienic gulf separating rich and poor (Tanner et al, 1983). At present it is virtually impossible to study maturation rates in earlier populations for want of a means of calibrating the stages, although comparison with dental development would doubtless provide interesting differences with modern populations. Illness and nutritional stress influence dental development less than linear growth. For this reason dental development can be used as a standard against which skeletal growth may be judged. The children at Poundbury Camp were small for their age using the criterion of dental age (Fig. 2).

Comparison of two methods of ageing rarely give good correlations. Deciding which ageing method is the more accurate requires a closer examination of other methods. Trabecular density and cortical remodelling have so far received little attention in this country. At first sight the ages derived from these methods do not match well with the dental ages. The fault may lie with the dental ages. Other methods of ageing are being explored. It has recently been demonstrated that teeth continue to erupt throughout life, an observation implicit in the expression 'getting long in the tooth'. The degree of over-eruption has been measured by Whittaker et al (1984a) and Levers and Darling (1983). It should be possible to calibrate the rate of eruption for a given population and so provide an ageing scale. The development of degenerative changes in the spine has been documented by Sager (1969). The rate of development of these changes in the Poundbury population is currently being studied by Sarah Atkinson. Preliminary results show that the rate of change is different in males and females.

Sex differences

Sex differences are most apparent in the developmental processes but are also important in some of the degenerative changes. It follows that wherever possible the sex should be determined and males and females scaled and calibrated separately. Demirjian and Levesque (1980) demonstrated sex differences in tooth development. For the first stages of crown formation of the majority of permanent teeth there is no difference in the chronology of dental calcification between boys and girls. For the completion of crown development, girls are more advanced than are boys by an average of 0.35 years for four teeth. For the following stages of root development the mean difference between sexes for all teeth is 0.54 year. The canine in particular is delayed in boys.

Girls are more skeletally mature than boys from birth onwards (Tanner 1962). This is true in all populations but until there is a reliable method of sexing the bulk of juvenile material we cannot exploit this fact in the study of skeletal material of the past and age determination of juveniles based on skeletal maturity criteria must defer to this irresolution. Thus, if the bone appearances are those of either a girl of eight or a boy of ten then the total range of the age estimate is between six years and 12 years. Further differences between the sexes that are related to ossification occurring around puberty are surprisingly poorly documented - not because the studies have not been carried out but because of the great variation in results presented. All the age indicators used on adult skeletons are known to progress at different rates in males and females. The reasons for this may be both intrinsic and environmental.

Reliability of Methods

It is a frequent observation that there is a wide margin of error in the assessment of skeletal age both between workers and by one worker at different times. This applies equally to radiographic studies of the living where age can be ascertained and to skeletal studies. Anthropologists have yet to agree on the significance of ageing indicators. Masset (1976) cites as an example that a given pubic symphysis would be aged at 36 years according to the method of McKern and Stewart (1957) and 58 years according to Nemeskeri et al (1960).

It is becoming clear that certain methods habitually underestimate age at death for at least part of the range and for some populations, while other methods tend to overestimate. The Recommendations for Age and Sex Diagnosis of Skeletons drawn up by the Workshop of European Anthropologists in 1978 and published in 1980 recognises that because of the strong individual variability in age changes in the skeleton, it will never be possible to determine with exact certainty the age of death of skeletons as can be done for recent populations from written documents. Similarly the Workshop felt that it was not possible to determine the differences in the tempo of maturation and ageing processes in different populations. There are ethnic and diachronic differences in the tempo of the ontogenetic development (Eveleth & Tanner, 1976; Legoux, 1962; Garn et al, 1964), in particular the acceleration of growth due to improved nutrition.

I think that the Workshop was unduly pessimistic on this point. Dental development is less affected by nutritional or disease stress than is skeletal development (Boas, 1933; Trowell et al, 1954) and can be used as a datum for

the estimation of skeletal development (Plate 2). There are several studies that indicate that children in Romano-British and mediaeval times were two to four years behind their modern equivalents in skeletal size (Figure 2). It is interesting to note that children bearing the marks of previous stress in the form of lines of arrested growth seen on radiographs of the tibia, are not shorter than their unmarked contemporaries. Catch-up growth on recovery is rapid and apparently complete (Mays, 1985). The Workshop urges uniformity of practice among anthropologists, so that results from different populations can be compared. The standards given for size, however, are derived from very small samples of a medieval Slavonic and a southern German population. A modern standard such as that of Maresh (1955) might have been preferable and no more inappropriate for a pre-Anglo-Saxon population of England. The complex method of Nemeskeri et al (1960) with the impracticality and expense of sectioning large numbers of bones precludes the widespread acceptance of their methods at least until they have been better evaluated.

The authors of the Recommendations are aware of their limitations, noting that there are sex differences in the development of the spongiosa of the femur head and the influence of births on age changes of the pubic symphysis of women. There are also differences in the ageing process depending on heredity and environmental factors such as nutrition, disease and workload. It is rather surprising that dental attrition, so widely used by anthropologists and archaeologists in Britain, received such scant attention from the Workshop. The method has its faults, tending to underestimate age at death, but it is probably no worse than the other methods of ageing recommended. Once the attrition grades have been re-evaluated and a more realistic scale devised the method could be just as reliable as any other (Figure 3).

PALAEODEMOGRAPHY

The credibility of the methods of ageing skeletal materials must be tested when we come to examine the distribution of dead in each age category. For most cemeteries the age frequency distribution bears little resemblance to any known population from a modern, developed or undeveloped, or historical source. We seldom, surely too seldom, record death over forty. This together with the evidence from the incidence of late age of onset disease like osteitis deformans and hyperostosis frontalis interna (Plate 3) points to a fault in our ability to identify older individuals. The record for earlier years is probably reasonably accurate. There is a dearth of adolescents in cemetery material;

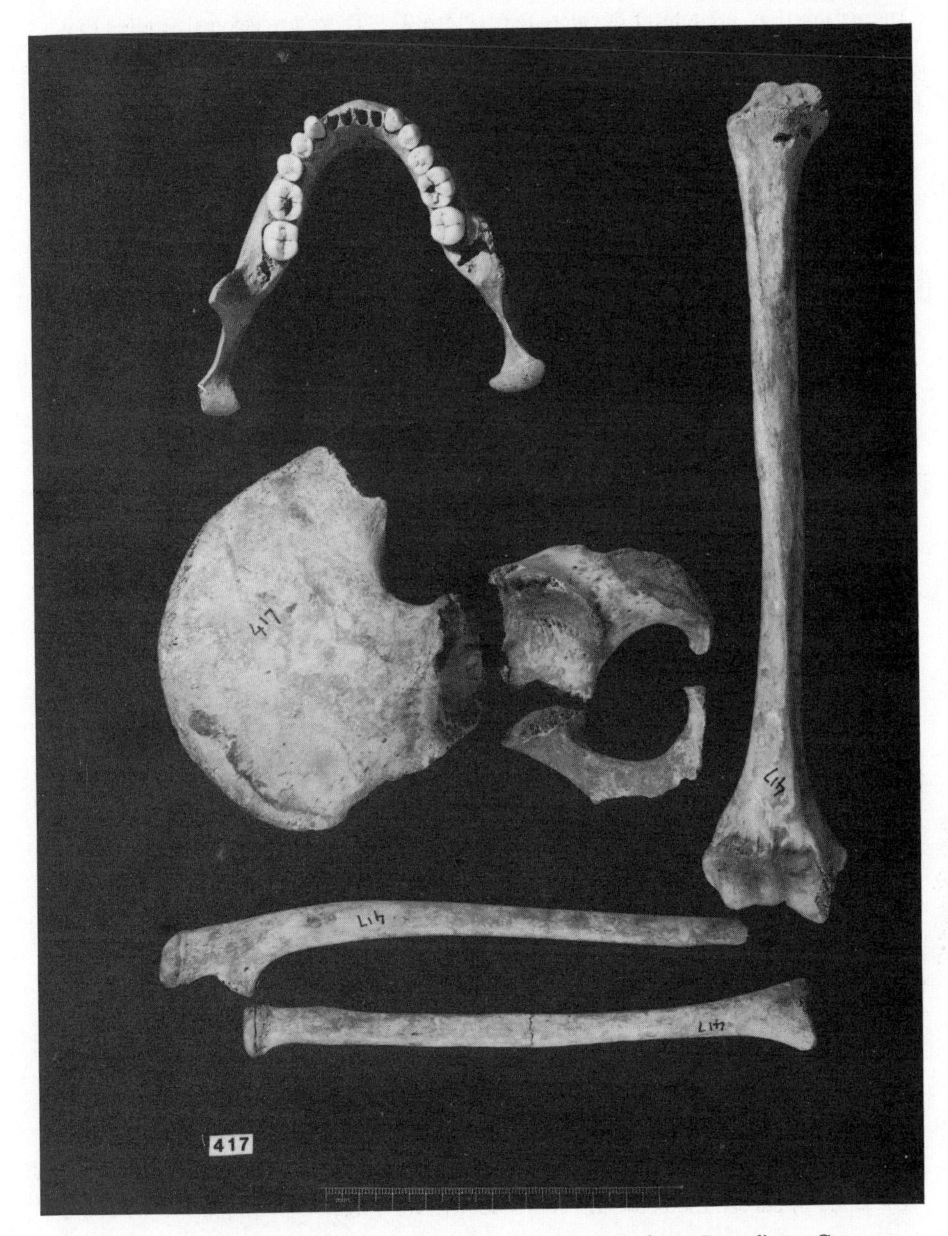

Plate 2. Conflicting ageing information in a juvenile from Poundbury Camp. The 2nd molar has erupted (9-14 years), the pubis and ischium have not fused (under 6 years), a secondary centre is present in the acetabulum (over 12 years). The epiphyses of the elbow, except the medial epicondyle, are fusing (under 18 years).

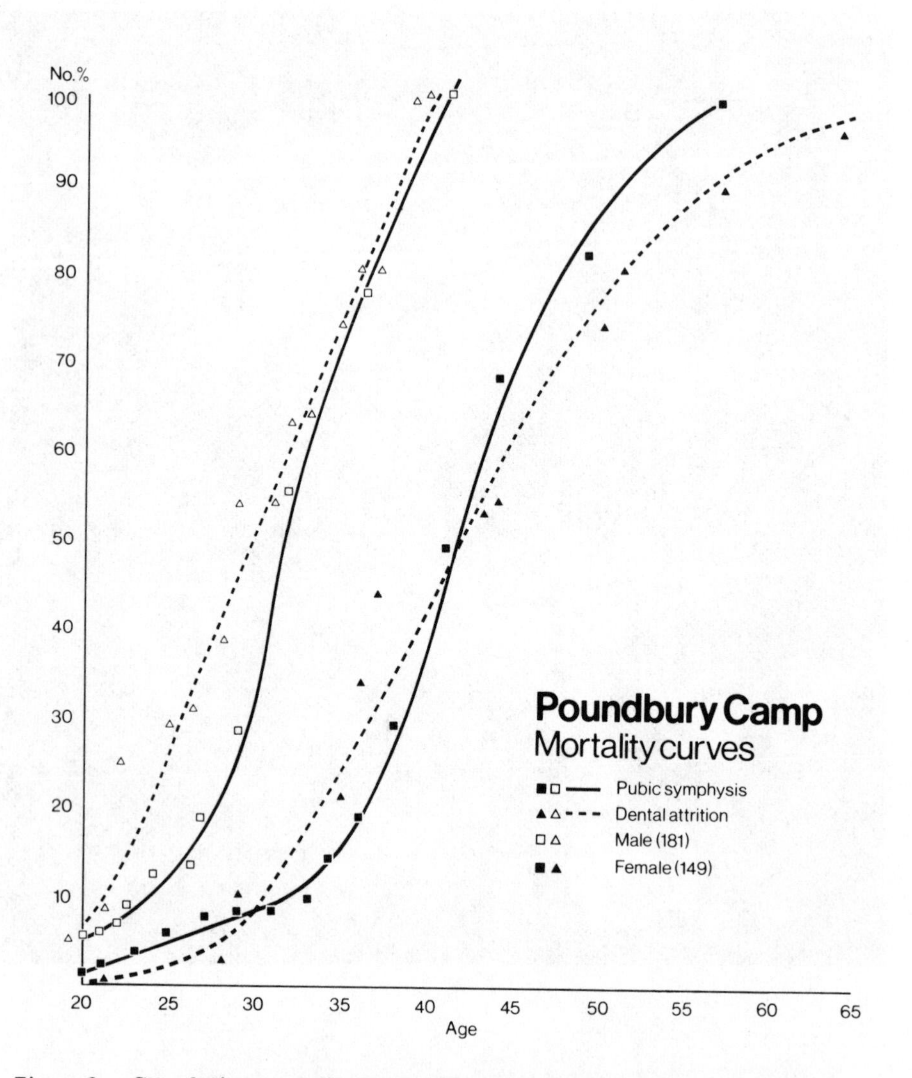

Figure 3: Cumulative mortality curves for males and females from Poundbury Camp based on two different and independent methods of ageing. The pubic symphysis wear was scored on a 15-point scale which has been calibrated using a known age modern sample (McKern & Stewart, 1957; Gilbert & McKern, 1973). The dental attrition was scored on a nine point scale (Brothwell, 1963) then calibrated taking into account attrition rates for the population. The agree-ment between the two the two methods is reasonably good. There does appear to be a real difference in mortality patterns between the two sexes. Dental attrition may be a more sensitive method for detecting mortality among young adults. The lack of males over 40 is in part but not wholly, an artefact of the method.

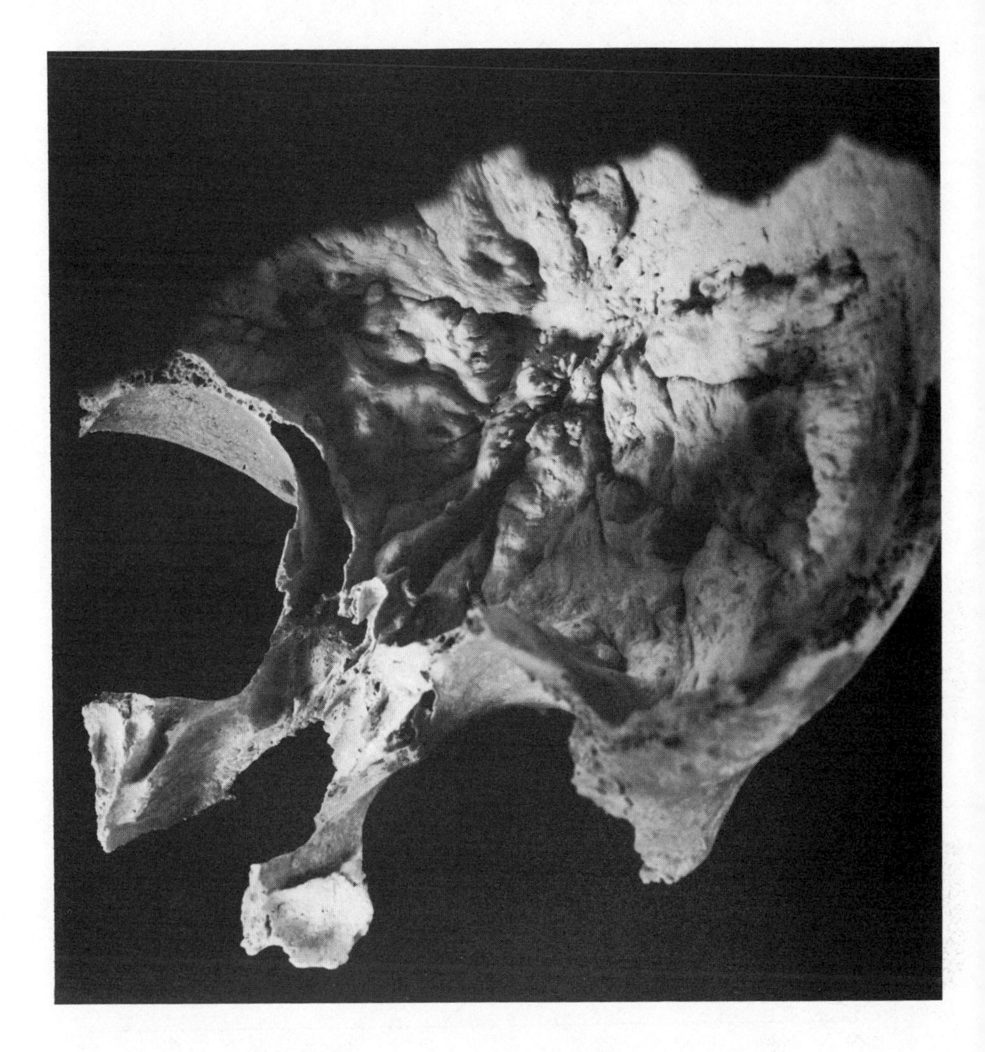

Plate 3: Hyperostosis frontalis interna - a gross thickening of the inner tabl of the frontal bone of the skull. This condition has an overall incidence of 4% but is more prevalent in obese, hirsute women over 45. There are about 20 cases among the 1430 skeletons at Poundbury Camp (PC 75E 886).

peak ages of death for children are observed between 2 and 4 and 6 and 8 years. The rates for young adults, especially females, probably do reflect real mortality at these ages. The causes of these varying death rates are not clear. Two to four year olds are being weaned, become susceptible to infectious disease, and are beginning to toddle and explore the world about them. A peak at 6-8 years suggests some more social cause, perhaps when the children were left to fend for themselves. The high mortality for women in their late teens and early twenties may be attributable to pregnancy and childbirth but this may be too facile an explanation. The high rate of female mortality is not maintained in later years and we find similar numbers of males and females in the middle age category (Molleson, 1981).

Male mortality is at its highest in the late twenties and early thirties even in the non-warring people buried at Poundbury. This may reflect a larger sample within the cohort. Poundbury Camp was one of the cemeteries outside the walls of the Roman town of Dorchester (Durnovaria). The large number of males in the 25-35 years age category may be a consequence of the immigration of young men to the town with its opportunites for increased wealth and power. Such immigrants, if they came from the less densely populated surrounding countryside, would have been especially vulnerable to infectious disease.

Bocquet and Masset (1977) have devised a method of estimating mathematically some demographic parameters, even for very old cemeteries. Life expectancy at birth, infant and child mortality, and birth and mortality rates were modelled. When these predictions were tested on a number of cemetery populations where, exceptionally, all the infants appear to have been preserved, the results of the predictions came close to the observed data in the case of two Mesolithic sites from North Africa but indicated that the number of infants (under one year) was quite inordinate in the case of the Greek Neolithic site of Lerna (Angel, 1971). The age structure of a cemetery population may be, and probably is, quite different from the age structure of the living population from which it derived. The number dying at a given age is a function of the death rate acting on the number of that age available in the population. The number available in the population is dependent on a variety of factors only one of which is birth rate. Other factors are neonate mortality, immigration and emigration.

Birth rate and population size cannot be inferred from a cemetery population as has been pointed out by Sattenspiel and Harpending (1983). If the effective birth rate increases, then the mean age at death increases. Conversely an increase in mean age at death implies an increase in birth rate, not an increase

in life expectancy at birth. But an increase in mean age at death can also be due to immigration. In a stationary population the mean age at death can be used to estimate the average length of life in the population. Such populations are unlikely in reality but are approached in cemeteries with a time depth of many generations. In order to calculate the expectation of life at birth in a non-stationary population it is necessary to have some estimate of the total size of each living cohort. These data are not available at all from the archaeological record and even estimates of total population size are generally unsatisfactory.

Reconstruction of population structure is therefore in most instances virtually an impossible task. The death assemblage can tell us very little of the living population. It is often distorted by such burial practices as exclusion of most newborn infants or even unnamed children. Among hunters, men may be buried where they made their first killing. Suicides, lepers and felons have all been excluded from community cemeteries in the past. These groups were rarely numerous and were often outcasts in their lifetime. A more significant group often excluded from the community would be the soldiers killed in battle and buried abroad. Their absence would hardly be detected in our cemetery assemblages although healed battle injuries among the older individuals would be a clue.

Exceptionally, as at Poundbury Camp, the burial assemblage probably is representative of the deaths that occurred in an early Christian community. There is no evidence for the sort of injuries that are sustained in warfare. Males and females are more or less equally represented and differences in sex ratios at different ages probably are meaningful. The high mortality among young women is most probably related to childbearing. Many of these women show no sign of having borne a child. We can speculate that they miscarried and it is tempting to speculate further that this is related to the high levels of lead detected in the bones and teeth of the skeletons (Waldron et al, 1979; Whittaker & Stack, 1984). It is beyond speculation to account for the lack of males over 45 years old (Figure 3), or indeed for the generally higher mortality among males, or is it decelerated ageing?

REFERENCES

Acsadi, G. & Nemeskeri, J. (1970). History of Human Life Span and Mortality. Budapest: Akademiai Kiado.

Angel, J.L. (1971). Lerna 11: The People. Washington: D.C. Smithsonian Institution.

Boas, F. (1933). Studies in Growth II. Human Biology, **5**, 429-444.

Bocquet, J-P. & Masset, C. (1977). Estimateurs en Paleodemographie. L'Homme, **17**, 65-90.

Brothwell, D.R. (1963). Digging up Bones. British Museum (Natural History), 1st edn.

Brown, W.A.B., Molleson, T.I. & Chinn, S. (1984). Enlargement of the frontal sinus. Annals of Human Biology, **11**, 221-226.

Comas, J. (1960). Manual of Physical Anthropology. Springfield: Charles Thomas.

Demirjian, A., Goldstein, H. & Tanner, J.M. (1973). A new system of dental age assessment. Human Biology, **45**, 211-227.

Demirjian, A. & Levesque, G-Y. (1980). Sexual differences in dental development and prediction of emergence. Journal of Dental Research, **59**, 1110-1122.

Eveleth, P.B. & Tanner, J.M. (1976). Worldwide variation in human growth. Cambridge University Press.

Fazekas, I.G. & Kosa, F. (1978). Forensic Fetal Osteology. Budapest: Akademiai Kiado.

Garn, S.M., Silverman, F.N. & Rohmann, C.G. (1964). A rational approach to the assessment of skeletal maturation. Annals of Radiology, **7**, 297-307.

Gilbert, B.H. & McKern, T.W. (1973). A method for ageing the female os pubis. American Journal of Physical Anthropology, **38**, 31-38.

Greulich, W.W. & Pyle, S.I. (1959). Radiographic Atlas of Skeletal Development of the Hand and Wrist, 2nd edn. Stanford: Stanford University Press.

Gustafson, G. (1950). Age determinations on teeth. Journal of the American Dental Association, **41**, 45-54.

Hansen, G. (1953). Die Alberbestimmung am proximalen Humerus-und Femurendi im Rahmen der Identifizierung menschlicher skelettreste. Wiss.Z.Humboldt-Universitat, Berlin, Math. naturwiss, **3**, 1-73.

Harrison, R.J. (1958). Man the Peculiar Animal. Harmondsworth: Pelican.

Johanson, G. (1971). Age determinations from human teeth. Odont.Revy. **22**, Supplement 21.

Krogman, W.M. (1962). The Human Skeleton in Forensic Medicine. Springfield: Charles Thomas.

Legoux, P. (1962). Determination de l'age dentaire de quelques fossiles de la ligne humaine. Rev.franc.Odonto.Stomat. **9**, 1165-1214, 1317-1330; **10**, 1031-1048.

Levers, B.G.H. & Darling, A.I. (1983). Continuous eruption of some adult human teeth in ancient populations. Archives of Oral Biology, **28**, 401-408.

Lunt, R.C. & Law, D.B. (1974). A review of the chronology of calcification of deciduous teeth. Journal of the American Dental Association, **89**, 599-606.

Manouvrier, L. (1984). Sutures. Dictionnaire des Sciences Anthropologiques.

Maresh, M.M. (1940). Paranasal sinuses from birth to late adolescence. American Journal of Diseases in Children, **60**, 55-78.

Maresh, M.M. (1955). Linear growth of long bones of extremities from infancy through adolescence. American Journal of Diseases in Children, **89**, 725-742.

Martin, R. & Saller, K. (1957). Lehrbuck der Anthropologie. 3 Aufl. Stuttgart.

Masset, C. (1976). Sur quelques facheuses methode de determination de l'age des squelettes. Bull.Mem.Soc.Anthrop.Paris. **3**, 329-336.

Mays, S. A. (1985). The relationship between Harris line formation and bone growth and development. Journal of Archaeological Science, **12**, 207-220.

McKern, T.W. & Stewart, T.D. (1957). Skeletal age changes in young American males, analysed from the standpoint of identification. Headquarters Quartermasters' Research & Development Command. Tech.Rep.EP-45. Natick, Massachusetts.

Miles, A.E.W. (1963). The dentition in the assessment of individual age in skeletal material, In: D.R. Brothwell (ed.), Dental Anthropology. Oxford: Pergamon Press.

Molleson, T.I. (1981). The archaeology and anthropology of death: what the bones tell us, In: S.H. Humphries (ed.), Mortality and Immortality, pp. 15-32. London: Academic Press.

Morrees, C.F.A., Fanning, E.A. & Hunt, E.E. Jr. (1963). Age variation of formation stages for ten permanent teeth. Journal of Dental Research, **42**, 1490-1502.

Nemeskeri, J., Harsanyi, L. & Acsadi, G. (1960). Methoden zur Diagnose des Lebensalters von Skelettfunden. Anthrop.Anz. **24**, 70-95.

Sager, P. (1969). Spondylosis cervicalis. A pathological and osteoarchaeological study. Copenhagen: Munksgaard.

Sattenspiel, L. & Harpending, H. (1983). Stable populations and skeletal age. American Antiquity, **48**, 489-498.

Scheuer, J.L., Musgrave, J.H. & Evans, S.P. (1980). The estimation of late fetal and perinatal age from limb bone length by linear and logarithmic regression. Annals of Human Biology, **7**, 257-265.

Schranz, D. (1933). Der Oberarmknocken und seine gerichtlich-medizinische Bedeutung aus dem Gesichtspunkte der Indentiat. Deutsch.Zeitschr. ges.ger.Med. **22**, 332-361.

Sjovold, T. (1975) Tables of the combined method for determination of age at death given by Nemeskeri, Harsanyi and Acsadi. Anthrop.Kozl. **19**, 9-22.

Spencer, R.P. & Coulombe, M.J. (1966). Quantitation of the radiographically determined age dependence of bone thickness. Investigative Radiology, **1**, 144.

Stack, M.V. (1967). Vertical growth rates of the deciduous teeth. Journal of Dental Research. **5**(Suppl.), 879-882.

Stack, M.V. & Halazonetis, J. (1973). Velocity of root growth in the central incisors of the deciduous dentition. Anatypon Aristotelian pampistemion Thessalonekis, 67-71.

Stewart, T.D. (1958). The rate of development of vertebral osteoarthritis in American whites and its significance in skeletal age identification. The Leech, **28**, 144-151.

Stloukal, M., Vyhnanek, L. & Rosing, F.W. (1957a). Spondylosehaufigkeit bei mittelalterlichen Populationen. Homo, **21**, 47-53.

Stloukal, M. & Vynanek, L. (1975b). Die Arthrose der grossen Gelenke. Homo, **26**, 121-136.

Stloukal, M. & Hanakova, H. (1978). Die Lange der Langsknocken alt slawischer Bevolkerungen - Unter besonderer Berucksichtigung von Wachstumfragen. Homo, **29**, 53-69.

Sundick, R.I. (1978). Human skeletal growth and age determination. Homo, **29**, 228-249.

Tanner, J.M. (1962). Growth at Adolescence, 2nd edn. Oxford: Blackwell.

Tanner, J.M., Whitehouse, R.H. & Healy, M.J.R. (1962). A new system for estimating the maturity of the hand and wrist with standards derived from 2600 healthy British children. II. The scoring system. Paris: International Children's Centre.

Tanner, J.M., Whitehouse, R.H., Marshall, W.A., Healy, M.J.R. & Goldstein, H. (1975). Assessment of skeletal maturity and prediction of adult height: TW2 method. New York/London: Academic Press.

Todd, T.W. (1920). Age changes in the pubic bone. I: The male white pubies. American Journal of Physical Anthropology, **3**, 285-334.

Todd, T.W. (1921). Age changes in the pubic bone. VI: The interpretation of variations in the symphysial area. American Journal of Physical Anthropology, **4**, 407-424.

Todd, T.W. (1930). Age changes in the pubic bone. VIII: Roentgenografic differentiation. American Journal of Physical Anthropology, **14**, 255-272.

Trowell, H.C., Davies, J.N.P. & Dean, R.F.A. (1954). Kwashiorkor. London: Arnold & Co.

Virtama, P. & Helela, T. (1969). Radiographic measurements of cortical bone. Acta Radiologica, Supplementum, **293**, 268.

Vlcek, E. & Mrklas, L. (1975). Modification of the Gustafson Method of Determination of Age According to teeth on prehistorical and historical Osteological Material. Scripta medica, **48**, 203-208.

Waldron, H.A., Khera, A., Walker, G., Wibberley, G. & Green, C.J.S. (1979). Lead concentration in bones and soil. Journal of Archaeological Science, **6**, 295-298.

Whittaker, D.K. & Stack, M. (1984). The lead, cadium and zinc content of some Romano-British teeth. Archaeometry, **26**, 37-42.

Whittaker, D.K., Molleson, T., Daniel, A.T., Williams, J.T., Rose, P. & Resteghini, R. (1985a). Quantitative assessment of tooth wear, alveolar crest height and continuing eruption in a Romano-British population. Archives of Oral Biology, **30**, 493-501.

Whittaker, D.K., Davies, G. & Brown, M. (1985b). Tooth loss, attrition and temporomandibular joint changes in a Romano-British population. Journal of Oral Rehabilitation (in press).

Workshop of European Anthropologists (1980). Recommendations for age and sex diagnoses of skeletons. Journal of Human Evolution, **9**, 517-549.

Zuhrt, R. (1955). Stomatologische Untersuchungen an Spatmittelalterlichen Funden von Reckkahn (12-14 jh). I. Die Zahnkaries und ihre Folgen. Dtsche. Zahn. Mund. und Kieferheilkunds, **25**, 1-15.

CELL DEATH AND
THE LOSS OF STRUCTURAL UNITS OF ORGANS

D. BELLAMY
Department of Zoology,
University College, Cardiff, U.K.

The meaning of the term 'ageing' is open to much dispute when applied to biological studies (Chandler, 1952). Some workers use the term to cover all aspects of development, from fertilization to death, i.e. it stands simply for an increase in years; others restrict the ageing phase to the period when there is a rapid decrease in life expectancy. For the purposes of the present discussion ageing is defined as "a decrease in adaptation as a consequence of loss of tissue and functional reserves" (Albertini, 1952). This definition implies that a major feature of ageing is a decline in the ability of the homeostatic systems of the body to cope with fluctuations in the external world. An important qualification is that Albertini's external world is the protective one that we have created through social evolution for ourselves and our valued livestock. In the 'wild', homeostatic failures would probably emerge before the equivalent of 'three-score years and ten' and be different from those of 'old age'.

Specific tissue and functional reserves decline during early development by deletion and remodelling of organs through adaptive processes which aid survival. The purpose of this article is to raise the questions: Are any of the cellular changes in post-embryonic tissues adaptive in this embryonic sense? Further, do they become non-adaptive, according to Albertini's definition, only when the species is allowed to live on, protected from the forces of selection which controlled its evolution? In evolution, deletion of structures, either through failure to detect and repair errors or through the active dismantling of functional components would be adaptive if it were to be connected with an advantageous diversion of the resources elsewhere in the body. In this connection, the body of an organism in its evolutionary environment would be a balance of defects, according to the relative advantages of approaching perfection in the metabolic specification of its various organ cell populations.

The principle of partition underlying this competition for resources is embodied in the empirical concept of allometric growth. Frequently, embryonic

growth curves can be approximated by simple exponential curves. If plotted semi-logarithmically, breaks in the straight lines occur representing the starting points of different rates of exponential growth, which are often timed differently for the various developing organs. These relationships are brought out more clearly when the rate of any physiological process is plotted logarithmically against another variable to give a resultant straight line. This is the so-called allometric relationship where the specific growth rate of a component or process, (y), stands in a constant ratio to the specific growth rate of either another component, or the total mass of the organism (x):

$$\frac{dy}{dt} \cdot \frac{1}{y} : \frac{dx}{dt} \cdot \frac{1}{x} = a$$

Physiologically, the mechanism underlying allometric growth becomes clear if this equation is written in a slightly different form:

$$\frac{dy}{dt} = a \cdot \frac{dx}{dt} \cdot \frac{y}{x}$$

That is, the cell population (y) receives from the increase in the resources entering the total system (dx/dt) a share which is proportional to its ratio to the total system (y/x). The allometric coefficient (a) is a distribution coefficient indicating the capacity of (y) to appropriate a certain share of the total increase. Allometric growth is therefore an expression of the competition of parts of the living organism for available resources. As in the embryo, breaks or jumps are frequently found in allometric plots which indicate the establishment of new priorities for the distribution of resources within the body. In this respect the major factor determining resource partition is size, not time. Functionally, therefore, allometry can be conceived of as being an expression of the principle that the various cell populations within the organism must remain in balance with the environmental resource despite variations of absolute size. So far, the principle of allometry has not been tested in relation to ageing except to establish the positive interspecific relationship between brain size and life span within the mammalian order. The experimental extensions of life spans by dietary restriction also stress the link between the rate of ageing and size. Unfortunately intraspecies relationships in error accumulations between the various organs have not yet been studied. This approach could reveal points during the growth phase at which certain organs begin to accumulate faulty structures, differentially, at a relatively rapid rate.

Several of the breaks in allometric plots coincide with body weights that are characteristic of times where changes in endocrine activity take place, particularly those concerned with sexual maturity. This, together with the obvious role of hormones in initiating large shifts in the use of metabolic resources, have led to ideas that ageing is bound up with primary alterations in the effectiveness of the endocrine system.

Hormones are integrated into the pattern of development in two ways. They act as controllers, in that they release or trigger the developmental potential inherent in certain tissues. Actions of this type are sometimes irreversible, but usually the target tissue returns to its former state when the concentration of the particular hormone falls to the initial level. Second, hormones invoke responses that offset undesirable changes in the external environment. From the latter point of view, the development of endocrine organs takes place in two stages. Initial development makes the animal more adaptable and independent of its surroundings. Later changes decrease the capacity for adaptation to the environment. The latter proposition may be discussed in terms of failures in the sensory detection of need, in the endocrine secretory mechanisms, in the clearance and metabolism of the circulating hormone and in the receptors of the target tissue and its functional capacity. There can be little doubt that endocrine homeostasis becomes less efficient in later life as a result of structural changes at all these points in the response cycle. Nevertheless, there are still those who propose that ageing is brought about largely by a single failure in a master endocrine clock located in the brain, which is driven by the amount of food eaten and its chemical composition.

If it is anticipated that a hormone deficiency is responsible for an ageing phenomenon, the obvious test is to treat the animal with the hormone in an attempt to restore the youthful characters. From many studies of this nature it appears that age differences in biochemistry, physiology and behaviour are not related solely to specific endocrine failures (Caldwell & Watson, 1954; Moon et al, 1956; Pauker et al, 1958; Everitt, 1959; Ackermann & Kheim, 1964; Rigby et al, 1984). The failure of hormones to maintain growth and prolong life, points to the great importance of other as yet unknown mechanisms for determining the species characteristics of body size and lifespan. From much of the early work concerned with defining the role of endocrine organs it is clear that the mammalian body is capable of functioning as a balance between competing cell populations without hormones, providing the external environment remains constant. In this respect, ageing appears to be bound up with the largely

unknown communication systems between cells in organs which ensure structural and functional integrity. This makes a multifactorial theory of ageing more likely than one based on a central clock idea (see Merry, this volume).

Albertini's view of ageing has been equated with 'senescence' (Medawar, 1981), but according to this definition, loss of adaptability begins long before there is a marked increase in mortality or decline in fertility. The fall in basal metabolism of man has been recognized for over a century, one of the best early studies being that of Sonden and Tigerstedt (1895). In terms of the loss of calories per unit area, the basal metabolic rate drops steadily until the age of 18, then more slowly thereafter (DuBois, 1936). The first phase is due to changes in the surface area to volume ratio connected with growth; the second phase has not been defined, but is expressed as an increased variability in metabolic rate at the highest body weights after growth has ceased. Although a general decline in human physiological performance is first noticeable during the third decade of life, this trend probably originates in events taking place in the late teens. The corresponding decline in the laboratory rat (life-span 800 days) occurs at about 200 days; in the fruit fly (Drosophila) ageing of some systems begins at about 10 days. If the resources devoted to mammalian 'play' are considered, there is an even earlier re-partition of energy consequent upon the adoption of the less vigorous behaviour patterns of maturity, which must reside in basic shifts in cellular function.

At this more basic level, loss of complexity, notably connected with the changes in the composition and relative amounts of connective tissue, start in the embryo. In actuarial terms, age 12 defines the point of maximum life expectancy. From this standpoint, loss of adaptability begins before sexual maturity, making it possible that some of this loss is not merely the outcome of late events of no evolutionary significance. In Medawar's terms, the early advantages of not devoting energy and materials to repair all errors with or without the withdrawal of support for complex structures of little survival value more than makes good the later disadvantages which this necessarily entails. This is particularly relevant since animals do not live long enough in the wild to disclose the possible fatal outcome.

The view that ageing is due to a general random deterioration in cells of all types is not supported by the biochemical, physiological and structural evidence. If the composition of the blood is taken as an indicator of the over-all state of homeostasis, there appears to be no impairment of functional capacity with age. The acid/base balance of blood in resting humans varies little

between the ages of 25 and 85 (Shock & Yiengst, 1950). There are no significant changes in the carbon dioxide tension, total carbon dioxide content and bicarbonate concentration. In agreement with this, the mean pH changes only from 7.400 to 7.268 between the second and eighth decades of life. Organic constituents of blood show a similar stability (Horvath, 1946; Praetorius, 1952; Rogers, 1951). The general constancy of the ionic composition of the body is borne out by studies on individual tissue and subcellular fractions (Griswold & Pace, 1956). No significant age changes are observed in total water, fat, potassium and sodium of liver, whilst the intracellular water, nitrogen, potassium and phosphorus in muscle decrease only between 5 and 8% (Yiengst et al, 1959). There is no change in muscle lactic acid and creatine (Horvath, 1946) and the analysis of human aortae (Kanabrocki et al, 1963) taken over an age range from 2 to 69 years, revealed no significant alterations in either total nitrogen or sulphur. Another study in subjects from 15 to 58 years indicated that aortic elastin and creatine also remained relatively constant (Myers & Lang, 1946).

A constancy of composition was also found to be a feature of the ribonucleic acid in ventricular muscle of the rat from 100 to 1200 days (Wulff et al, 1963). In keeping with this, the mean DNA per nucleus and the mean volume of nuclei of rat liver do not change over the age range of 12 to 27 months (Falzone et al, 1959).

The failure to detect age changes indicative of large scale failures in tissue function from what is really a widely ranging random sample of tissues and chemistry, also impressed the early workers using histological methods and prompted the view that organisms do not die from a lack of 'good' cells. Only moderate signs of deterioration are found in the histochemistry of liver and there are no obvious large-scale functional deficiencies at the structural level in brain. Kidney is marked by large structural alterations, but this appears as a decrease in the relative cell mass without a marked change in the nature of the cells that remain (Lowry et al, 1946).

Despite the evidence in favour of stability of tissue structure, age-changes in composition are observed as the exceptions to the rule. This applies to a small number of blood constituents (Ackermann & Kheim, 1964; Das, 1964). Often, age changes are not consistent between tissues: although there is no variation in the riboflavin content of human brain, heart and skeletal muscle over seven decades (Schaus & Kirk, 1956), aortic tissue is characterized by a 60% fall in its riboflavin content over the same period (Schaus et al, 1955). Also, in contrast to ventricular tissue, the foliar tissue

of the cerebellum loses 30% of its RNA throughout the life-span (Wulff et al, 1963). Some of the largest alterations in tissue composition have been found in the lipid fraction. Cholesterol in the blood of healthy women increases twofold in concentration from the ages of 20 to 60 (Swanson et al, 1955). There are also substantial increases in the concentration of tissue elastin, collagen (Hall et al, 1955; Schaub, 1963), mucopolysaccharides (Kirk & Dyrbye, 1955a,b) and calcium (Lansing et al, 1950; Yu & Blumenthal, 1963). These changes, although a general feature of many tissues are not universal (Streicher, 1958; Bashey et al, 1967; Sobel & Hewlett, 1967; Sobel et al, 1958).

Taking the available evidence on the gross composition of tissues, it appears that, on the whole, age changes are not very marked in those intracellular components such as water, nitrogen and inorganic ions which are concerned with fundamental cellular organization. On the other hand, there are marked changes in the chemistry of the extracellular compartment, notably in the proportion and composition of the ground substance. These changes are probably a reflection of the general tendency for there to be an age-dependent shift in the balance of cell populations. For example, the eosinophil count in rats decreases progressively with age, there being a 50% drop between 200 and 800 days. Despite this loss there is little change in other blood cells (Everitt & Webb, 1958). In ageing cattle, it has been observed that a decrease in blood lymphocytes is accompanied by a rise in the proportion of neutrophils and a fall in the total leucocyte count (Riegel & Nellor, 1966).

Cell loss in mammals is most marked in the thymus. Here the process which results in a progressive general decrease in total mass is accompanied by well-defined shifts in the balance between the various component cell types (Bellamy, 1984). Cellular involution is a feature also of skeletal muscle, there being a 30% loss of thigh muscle between the first and second year of life in male rats (Yiengst et al, 1959), although this change is not so marked in females. Taking total body potassium as a measure of cell mass the number of cells per unit body weight shows a steady decline in humans from the late teens to the age of 80 (Allen et al, 1960). In another study (Shock et al, 1963) it was found that the extracellular space did not change, although the total body water diminished significantly with increasing age. Intracellular water, calculated as the difference between total body water and extracellular space, showed a significant regression. This is interpreted as a reflection of the loss of functioning cells. From evidence of this nature it is now generally held that involution, or cell loss, is one of the most consistent features of ageing.

Theories of ageing have to explain not only the widespread disappearance of cells but also the abnormalities in structure in many of those remaining and the connection between these cellular abnormalities and the general failure in homeostasis and the onset of degenerative disease. So far there has been little success in this type of comprehensive analysis. Although it is generally accepted that ageing occurs through a failure to correct metabolic errors there is no unambiguous evidence as to how the errors occur. The literature presents a piece-meal approach to the problem according to the different specialists who attempt to find a solution. At the physiological level, an early effort to connect a failure in physiological regulation with a Gompertzian mortality curve utilized a model which involved measuring the blood loss necessary to kill rats of different ages (Simms, 1942). The exponential rise in the chances of death was related to a constant mean rate of increase in sensitivity to blood loss coupled with a steady increased variability between animals. At the other extreme of bodily organization, efforts continue to detect mis-specified proteins according to Orgel's "error catastrophe" theory.

Since cell loss appears to be the most obvious quantitative feature of ageing, it makes sense to enquire into the mechanism, particularly since cell death is an important feature of embryonic life where it results in re-modelling of organs and the deletion of non-adaptive cell populations. There can be little doubt in this connection, that cell death is part of a genetic programme of development. Does this type of programme continue into adult life, and could the withdrawal of surveillance systems and associated replacement mechanisms be part of a species-survival strategy?

Post-embryonic cell death is an obvious feature of the lymphoid organs (notably the thymus), skeletal muscle, brain and kidney. The loss of cellularity in the thymus affects mainly the lymphocytes and occurs through the operation of an internal 'clock-like' mechanism (Bellamy, 1984). Because of its early onset and completeness, it is difficult to see thymic involution other than as an evolved deletion programme for the removal of an 'expensive' cell population that has carried out its function. Indeed, in some mammals, the thymus begins to deteriorate in the embryo. Muscle ages by loss of its structural units, the fibres and motor-end-plates (Gutmann et al, 1968). In this respect, cellular death is not a haphazard affair. Brain also loses cells in an ordered fashion, cell death in the human brain being confined to well-defined areas, beginning in the second decade. There is also quantitative evidence for the selective loss of

peripheral nerves. For example, peripheral nerve degeneration appears to be responsible for the steady decline in the nasal plexis of cells that also begins in the second decade of life and which, eventually, probably results in the loss of sensory discrimination (Smith, 1942). At a physiological level the kidneys begin to deteriorate early and this can be measured in the third decade as a drop in clearance, indicative of a decrease in glomerular filtration, reduced effective plasma flow and a reduction in both excretory and resorptive Tm's. Part of this decreased renal performance may be ascribed to the anatomical narrowing and loss of blood vessels together with a persistent vasoconstriction. However, not all of the changes observed can be accounted for simply on the basis of a reduction in the number of vascular pathways and a structural narrowing of the vessels which remain, particularly in young people. Even the remaining vascular bed of old kidneys is capable of responding to vasodilators as effectively as the vessels in younger organs, which indicates that there is not a general vascular deterioration.

Renal involution may be measured by a decrease in kidney weight. In humans the kidneys reach their maximum size at about the age of 20 years. By the eighth decade, almost one third of this maximum weight has been lost. The weight change is accompanied by a fall in the number of glomeruli which amount to only two thirds to one half the early adult number (Moore, 1931). The epithelial cells of the tubules decrease in size, narrowing to a shortened thread-like structure with a reduction in coiling, so that thirty nephrons have been found in micro-dissected material in the space formerly filled by three (Loomis, 1936). Many pathologists over the years have attributed these changes to secondary responses to primary alterations in the renal arteries involving sclerotic changes in their walls. In this respect, it has been said that the sclerotic artery of the aged is as normal to that period of life as the thin and flexible vessel is to childhood. On the other hand, losses of glomeruli have been observed in ageing rats and dogs, following precisely the human pattern, where arteriosclerosis is uncommon. This kind of discussion highlights the inconclusive arguments that continue between pathologists and biologists with regard to the definition of 'normal' as opposed to 'pathological' ageing. The fact that not all glomeruli are reduced in size and that some tubules without glomeruli have a normal histological appearance is strongly indicative of a basic process of selective dismantling, perhaps overlaid by nutritional deficiencies caused by thickened inflexible blood vessels.

Broadly speaking, from this kind of analysis of a wide variety of data,

non-random cell loss appears to be the major factor leading to decreased structural complexity of organs. This generalisation also applies to the human skin epidermis. Here, cell loss is associated with the disappearance of its interdigitations with the dermis. These finger-like projections, sometimes called 'rete pegs', are arranged in juveniles to form a net, where the pegs are the lateral walls of each small, repeated, unit. With age, the highly ordered net-like structure is completely lost and the epidermis becomes a thin flat sheet. There is some evidence for regional differences in the rate of this change. With respect to the dermis, most work has been concerned with changes in ground substances, such as collagen and elastin, and this work has tended to obscure early work on the cellular changes. In this connection, one study on the discrete dermal arterio-venous anastomoses, the glomic units, showed that these characteristic cellular structures, which are thought to be concerned with temperature regulation, have been greatly reduced in number by the seventh decade (Popoff, 1934).

It is clearly important to study the relationships between cell loss, structure and disease, but the organ systems so far discussed do not offer the possibility of experimental manipulation in the human subject. What is needed is a human model which is available on biopsy, which has a complex functional unit-structure showing age-dependent degenerative disease, ideally open to drug treatment. All of these criteria are met by the gastric epithelium. In addition, appropriate animal models are available for studying healing and the actions of protective agents on the gastric epithelium (Robert et al, 1979; Mersereau & Hinchey, 1982).

The importance of degenerative disease of the digestive system is not in doubt. Cancer of the stomach, liver, gall bladder and bile ducts comprise almost one third of the total of all forms of cancer. Between 2 and 3% of all deaths are due to cancer of the stomach, liver and biliary passages. In addition, about 14% of all forms of cancer arise from the peritoneum, intestine, colon and rectum, comprising about 1% of all deaths. It has been said that at least half of all forms of fatal cancers arise in organs or tissues of the digestive system. All of these cancers show a rise in incidence with increasing age. This age incidence is also true of gastric and duodenal ulcers (Coggan et al, 1981; Langman, 1982), which through uncontrolled haemorrhage and perforation are responsible for an increased death rate, gastric ulcer giving rise to more deaths than duodenal ulcer. Since there are now several drugs available for successfully treating ulcers, which differ in the relative proportions of remissions, it is

convenient to consider gastric ulcer as a model to answer questions such as: Is degenerative disease connected with specific, structural failure in function of the gastric epithelium? Are ulcer remissions due to age-dependent failures in epithelial repair mechanisms? The ready availability of biopsies before and after drug treatment offers unique advantages in studying the associated human cell biology.

Considerable evidence is available concerning the relationship between gastric acidity after test-meals and age. At birth, the gastric mucosa of most mammals is differentiated sufficiently to secrete acid gastric juice and the concentration of both acid and pepsin gradually rises during development. In humans, they both reach a plateau by the late 20's. The mean acidity decreases from the late forties in males, falling to the lower level maintained in females from the early thirties (Vanzant et al, 1932; Segal et al, 1933), instances of achlorhydria being excluded. Analysis by volume shows that females secrete less than males and in both sexes the volume of acid secretion produced in response to histamine test meals decreases from the age of 20. Taking these two measurements together, there is a striking early decrease in total acid secretion with age, which occurs at about the same rate in both sexes (Polland, 1933). The incidence of achlorhydria also increases with age, ranging from about 4% in the third decade to about 30% in the seventh decade. Achlorhydria together with the decline in acid secretion, may well be connected with histological changes in the gastric epithelium.

From birth to maturity the mean free acidity increases about eightfold, whereas the surface area of the epithelium and its glandular bodies increase between ten- and twentyfold. In this respect, although there is as yet only qualitative evidence, it is likely that the functional units, the gastric pits, steadily increase in number and complexity. Later, the histology of the stomach epithelium undergoes profound changes, the various alterations being grouped under the term atrophic gastritis. These changes include a decrease in mucosal thickness, increases in leukocytic infiltration, lymphoid aggregates and intestinal metaplasia with de-differentiation of the fundic glands. These all begin to appear in the third and fourth decades and are present in between 80 to 90% of people by the fifth decade. In detail, the structural changes appear to consist of a decrease in the numbers of acid producing cells, a straightening and shortening of the gastric glands and the enlargement of the outer mucous-secreting cells (Bellamy et al, 1984). In principle, this situation may well be the same as that underlying ageing of kidney, namely a loss of functional units and the simplifica-

tion of those that remain. The great majority of gastric ulcers occur only in the pyloric gland area, proximal to the border zone between the fundic and pyloric gland (Oi et al, 1959). This area has a well-defined embryonic origin indicating that a specific weakness develops with age connected with a specific, localised pattern of embryonic cell proliferation and differentiation. In this connection, the stomach shows what many other localized degenerative changes demonstrate, namely a clear site specificity within an organ or vascular system.

Although as yet these histological changes cannot be connected directly with the increased incidence of gastric ulcer, it appears that the regenerated epithelium after treatment with anti-acid drugs is abnormal in the elderly, with large areas of 'frilly' goblet cells.

In summary, there is much circumstantial evidence suggesting that post-embryonic cell loss is not a random affair and is in part, at least, related to apparent failures to maintain the specific physiological units of organs, which become simplified or totally lost. This takes the problem of ageing close to the as yet unanswered questions of embryology connected with the genetics of cellular pattern. Does each morphological unit have a 'master cell' surveying and maintaining it, or does the entire cell population of a tubule, a neuromuscular junction or a gastric pit, exist through some kind of diffuse intercellular communication system involving all genes of the unit? It may turn out that the cellular deteriorations of these unit structures are simply due to accumulations of random errors in susceptible cell populations, but as yet the possible involvement of special mechanisms for making the best use of the body's resources has not been ruled out.

REFERENCES

Ackermann, P.G. & Kheim, T. (1964). The effect of testosterone on plasma amino acid levels in elderly individuals. Journal of Gerontology, **19**, 207-214.

Albertini, A. (1952). What is ageing? Journal of Gerontology, **7**, 452-463.

Allen, T.H., Anderson, E.C. & Langham, W.H. (1960). Total body potassium and gross body composition in relation to age. Journal of Gerontology, **15**, 348-357.

Bashey, R.I., Torrii, S. & Aangrist, A. (1967). Age related collagen and elastin content of human heart valves. Journal of Gerontology, **22**, 203-208.

Bellamy, D. (1984). Cell death. Symposium of the Society for Experimental Biology. Cambridge University Press, pp. 106-121.

Bellamy, D., Bradshaw, M.J. & Lewis, G.H.J. (1984). Gastric ulcer, a model for exploring the developmental origins of degenerative disease. Proceedings of the Society for Experimental Biology, Cardiff.

Caldwell, B.M. & Watson, R. (1954). An evaluation of psychological effects of sex hormone administration in aged women: II. Results of therapy after 18 months. Journal of Gerontology, **9**, 168-174.

Chandler, A.R. (1952). A note on the meaning of ageing. Journal of Gerontology, **7**, 437-438.

Coggan, D., Lambert, P. & Langman, M.J.S. (1981). 20 years of hospital admissions for peptic ulcer in England and Wales. The Lancet, i, 1302-1304.

Das, B.C. (1964). Indices of blood biochemistry in relation to age, height and weight. Gerontologia, **9**, 178-192.

DuBois, E.F. (1936). Basal Metabolism in Health and Disease. Philadelphia: Lea & Febiger.

Everitt, A.V. (1959). The effect of pituitary growth hormone on the ageing male rat. Journal of Gerontology, **14**, 415-424.

Everitt, A.V. & Webb, C. (1958). The blood picture of the aging male rat. Journal of Gerontology, **13**, 255-260.

Falzone, J.A., Barrows, C.H. & Shock, N.W. (1959). Age and polyploidy of rat liver nucleii as measured by volume and DNA content. Journal of Gerontology, **14**, 2-8.

Griswold, R.L. & Pace, N. (1956). A comparison of intracellular nitrogen and metal ion distribution patterns in the livers of young and old rats. Journal of Gerontology, **12**, 150-152.

Gutmann, E., Hanzlikova, V. & Jacoubek, B. (1968). Changes in the neuromuscular system during old age. Experimental Gerontology, **3**, 141-146.

Hall, D.A., Keech, M.K., Reed, R., Saxl, H., Tunbridge, R.E. & Wood, M.J. (1955). Collagen and elastin in connective tissue. Journal of Gerontology, **10**, 388-400.

Horvath, S.M. (1946). The influence of the aging process on the distribution of certain components of the blood and the gastrocnemius muscle of the albino rat. Journal of Gerontology, **1**, 213-223.

Kanabrocki, E.L., Fells, I.G., Decker, C.F. & Kaplan, E. (1963). Total hexosamine, sulphur and nitrogen levels in human aortae. Journal of Gerontology, **18**, 18-22.

Kirk, J.E. & Dyrbye, M. (1955a). The phenol sulphatase activity of aortic and pulmonary artery tissue in individuals of various ages. Journal of Gerontology, **11**, 129-133.

Kirk, J.E. & Dyrbye, M. (1955b). Hexosamine and acid-hydrolysable sulphate concentrations of the aorta and pulmonary artery in individuals of various ages. Journal of Gerontology, **11**, 273-281.

Langman, M.J.S. (1982). What is happening to peptic ulcer? British Medical Journal, **284**, 1063-1064.

Lansing, A.I., Rosenthal, T.B. & Alex, M. (1950). Significance of medial age changes in the human pulmonary artery. Journal of Gerontology, **5**, 211-215.

Loomis, D. (1936). Plastic studies in abnormal renal architecture, IV. Vascular and parenchymal changes in arteriosclerotic Brights disease. Archives of Pathology, **22**, 435-463.

Lowry, O.H., Hastings, A.B., McKay, C.M. & Brown, A.N. (1946). Histochemical changes associated with ageing, IV. Liver, brain and kidney in the rat. Journal of Gerontology, **1**, 345-357.

Medawar, P.B. (1981). The Uniqueness of the Individual. New York: Dover.

Mersereau, W.A. & Hinchey, E.J. (1982). Role of gastric mucosal folds in formation of focal ulcers in the rat. Surgery, **91**, 150-155.

Moon, H.D., Koneff, A.A., Li, C.H. & Simpson, M.E. (1956). Phaeochromocytomas of adrenals in male rats chemically injected with pituitary growth hormone. Proceedings of the Society for Experimental Biological Medicine, **93**, 74-77.

Moore, R.A. (1931). The total number of glomeruli in the normal human kidney. Anatomical Record, **48**, 153-168.

Myers, V.C. & Lang, W.C. (1946). Some chemical changes in the human thoracic aorta accompanying the ageing process. Journal of Gerontology, **1**, 441-444.

Oi, M., Oshida, K. & Sugirmura, S. (1959). The location of gastric ulcer. Gastroenterology, **36**, 45-56.

Pauker, J.D., Kheim, T., Mensh, I.N. & Kountz, W.B. (1958). Sex hormone replacement in the aged: I. Psychological and medical evaluation of administration of androgen-estrogen and androgen-estrogen combined with reserpine. Journal of Gerontology, **13**, 389-397.

Polland, W.S. (1933). Histamine test meals, an analysis of 988 consecutive tests. Archives of Internal Medicine, **51**, 903-919.

Popoff, N.W. (1934). The digital vascular system. Archives of Pathology, **18**, 295-330.

Prastorius, E. (1951). Plasma uric acid in aged and young persons. Journal of Gerontology, **6**, 135-137.

Riegel, G.D. & Nellor, J.E. (1966). Changes in blood cellular and protein components during ageing. Journal of Gerontology, **21**, 435-468.

Rigby, M.K., Soule, S.D., Barber, W. & Rothman, D. (1964). Sex hormone replacement in the aged. Journal of Gerontology, **19**, 313-316.

Robert, A., Nezamis, J.E., Lancaster, C. & Hanchar, A.J. (1979). Cytoprotection by prostaglandins in rats. Prevention of gastric necrosis produced by alcohol, HCl, NaOH, Hypertonic NaCl and thermal injury. Gastroenterology, **76**, 88-93.

Rogers, J.B. (1951). The ageing process in the guinea pig. Journal of Gerontology, **6**, 13-16.

Schaus, R. & Kirk, J.E. (1956). The riboflavin concentration of brain, heart and skeletal muscle in individuals of various ages. Journal of Gerontology, **11**, 147-150.

Schaus, R., Kirk, J.E. & Laursen, T.J.S. (1955). The riboflavin content of human aortic tissue. Journal of Gerontology, **10**, 170-177.

Segal, Z., Marks, J.A. & Kantor, J.L. (1933). The clinical significance of gastric anacidity. A study of 6679 cases with digestive symptoms. Annals of Internal Medicine, **7**, 76-88.

Shock, N.W. & Yiengst, M.J. (1950). Age changes in the acid-base equilibrium of the blood of males. Journal of Gerontology, **5**, 1-4.

Shock, N.W., Watkin, D.M., Yiengst, M.J., Norris, A.H., Gaffney, G.W., Gregerman, R.I. & Falzone, J.A. (1963). Age differences in the water content of the body as related to basal oxygen consumption in males. Journal of Gerontology, **18**, 1-8.

Simms, H.S. (1942). The use of a measurable cause of death (haemorrhage) for the evaluation of ageing. Journal of General Physiology, **26**, 169-178.

Smith, C.G. (1942). Age incidence of atrophy of olfactory nerves in man. Journal of Comparative Neurology, **77**, 589-595.

Sobel, H. & Jewlett, M.J. (1967). Effect of age on hyaluronic acid in hearts of dogs. Journal of Gerontology, **22**, 196-198.

Sobel, H., Hrubant, H.E. & Hewlett, M.J. (1968). Changes in the body composition of C57BL/aa mice with age. Journal of Gerontology, **23**, 387-389.

Sonden, K. & Tigerstedt, R. (1895). Die Respiration und der gestammstoffwechel des Menshen. Skandinavanian Archives of Physiology, **6**, 1-224.

Streicher, E. (1958). Age changes in the calcium content of rat brain. Journal of Gerontology, **13**, 356-358.

Swanson, P., Leverton, LR., Gram, M.R., Roberts, H. & Pesek, I. (1955). Blood values of women: cholesterol. Journal of Gerontology, **10**, 41-47.

Vanzant, F.R., Avarez, W.C., Eusterman, G.B., Dunn, H.L. & Berkson, J. (1932). The normal range of gastric acidity from youth to old age. Archives of Internal Medicine. **49**, 345-359.

Wulff, V.J., Piekielniak, M. & Wayner, M.J. (1963). The ribonucleic acid content of tissues of rats of different ages. Journal of Gerontology, **18**, 322-325.

Yiengst, M.J., Barrows, C.H. & Shock, N.W. (1959). Age changes in the chemical composition of muscle and liver in the rat. Journal of Gerontology, **14**, 400-404.

THE PROSPECTS FOR MORTALITY DECLINE AND CONSEQUENT CHANGES IN AGE STRUCTURE OF THE POPULATION

B. BENJAMIN

Centre for Research in Insurance and Investment,
The City University, London, U.K.

INTRODUCTION

Over the past 150 years there have been profound changes in the sex and age structure of the United Kingdom as, first, a rise in the size of birth generations made the age structure very young as compared with a stationary population supported by constant births and experiencing constant mortality and, later, a fall in the annual numbers of births produced an ageing of the population structure. This ageing has been accentuated by steady mortality improvement, more so for women than for men. Small contributions to the ageing of the population structure were made by the virtual cessation of any substantial emigration of young men after the world economic depression of 1930 and by the war losses of young men during 1914-1918 and during 1939-1945. All countries of advanced or advancing economic development have suffered these changes and to the extent that many have in their economic development leap-frogged over the relatively slower industrial revolution of the United Kingdom, their changes in age structure have been more rapid and, in their economic implications, more violent. It is not the purpose of this presentation to make international demographic comparisons but it is important to remember that any economic problems arising from the growth in the dimensions of elderly dependency are common to all developed countries and will sooner or later confront the developing countries.

THE PAST CHANGES IN POPULATION STRUCTURE

Table 1 shows the structure of the United Kingdom at the Population Censuses 1871, 1901 and 1981. The changes are shown more dramatically by the population pyramids of Figure 1.

Table 1. The Population of the United Kingdom in 1871, 1901, 1981 (in thousands)

		1871		1901		1981	
		No.	%	No.	%	No.	%
Males	0- 4	1,850	6.74	2,190	5.73	1,766	3.14
	5-14	3,124	11.39	4,024	10.52	4,165	7.40
	15-24	2,460	8.97	3,636	9.51	4,510	8.02
	25-44	3,347	12.20	5,153	13.48	7,419	13.19
	45-59	1,560	5.69	2,214	5.79	4,757	8.46
	60-64	363	1.32	489	1.28	1,398	2.48
	65-74	435	1.58	565	1.48	2,267	4.03
	75 & over	169	0.62	219	0.57	1,062	1.89
	TOTAL	13,308	48.51	18,490	48.36	27,344	48.61
Females	0- 4	1,841	6.71	2,190	5.73	1,675	2.98
	5-14	3,094	11.28	4,016	10.50	3,954	7.03
	15-24	2,611	9.52	3,865	10.11	4,347	7.73
	25-44	3,735	13.61	5,647	14.77	7,322	13.01
	45-59	1,701	6.20	2,424	6.34	4,875	8.67
	60-64	412	1.50	577	1.51	1,583	2.81
	65-74	509	1.86	712	1.86	2,948	5.24
	75 & over	221	0.81	312	0.82	2,204	3.92
	TOTAL	14,124	51.49	19,743	51.64	28,908	51.39
GRAND TOTAL		27,432	100.00	38,233	100.00	56,252	100.00

Notes:

United Kingdom includes England, Wales, Scotland and Northern Ireland;
Percentage is percentage of grand total;
Sources: 1871, 1901 - Registrar General - Census figures. 1981 - mid-year, as adopted by Government Actuary for projection purposes.

It can be seen that in 1871 after many decades of a rising flow of births the population structure of the United Kingdom was extremely youthful. With the prevailing mortality rates a stationary population deriving from a constant size of generations of birth would have 26.36 per cent of persons under age 15 and 8.38 per cent over age 65. In fact the respective percentages were 36.12 and 4.87. The actual population pyramid of Figure 1 for 1871 has a wide base and a rapid thinning down to a narrow top. In 1901 after the birth flow (annual numbers of births) had been declining slowly, the structure had become slightly less youthful; the comparable percentages were 32.48 and 4.73. The change in the shape of the population pyramid is hardly noticeable. But soon after the turn of the century the birth rate fell sharply, partly as a consequence of an equally sharp decline in fetal and infant mortality so that there was less birth wastage

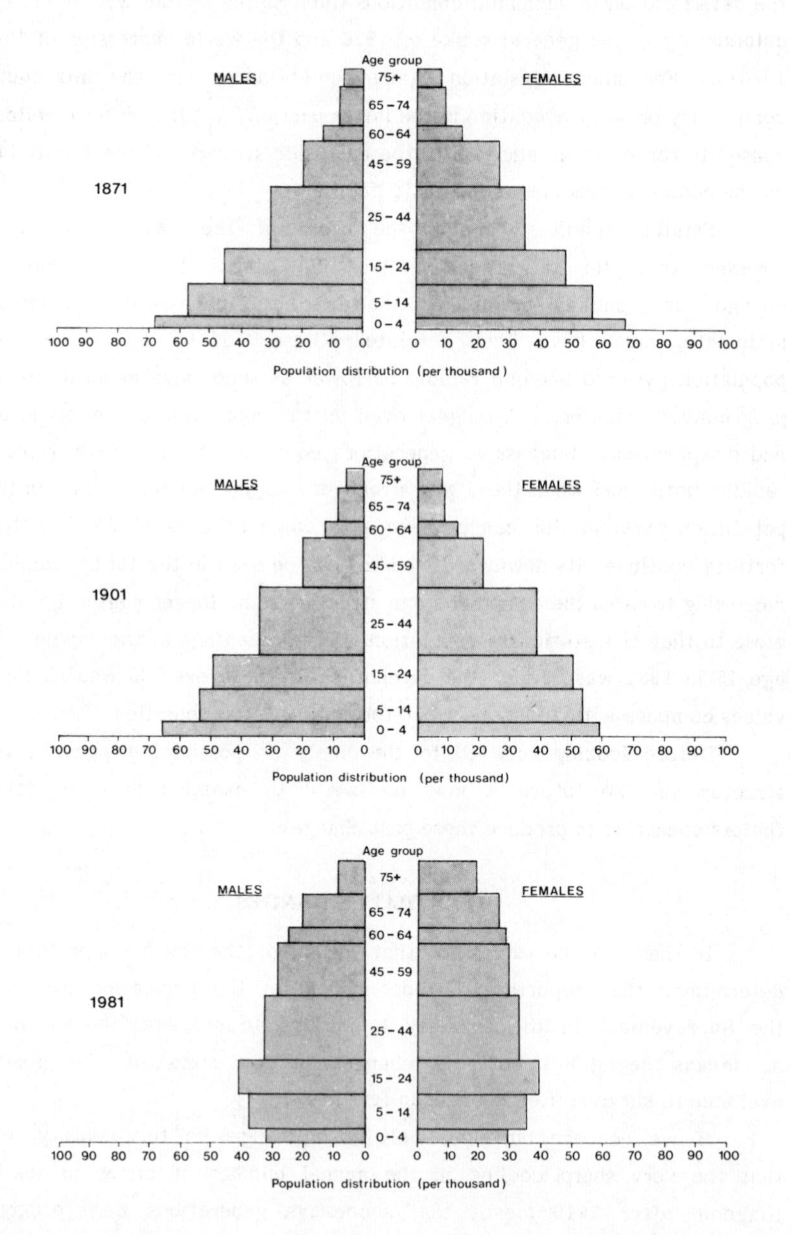

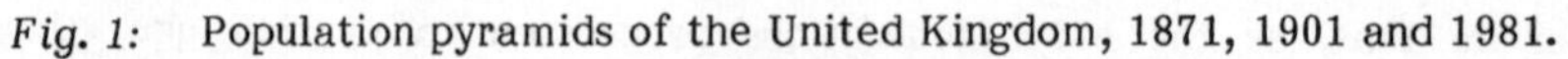
Fig. 1: Population pyramids of the United Kingdom, 1871, 1901 and 1981.

to overcome to produce a family of a desired size and partly as a consequence of the deterioration in economic conditions (interrupted by the war of 1914-18 but culminating in the general strike of 1926 and the world depression of the early 1930's). For since legislation against child labour and the introduction of compulsory primary education in the latter part of the 19th century, children had ceased to represent an addition to the economic strength of the family but had, on the contrary, become an economic liability.

Smaller families became the order of the day. It has been stressed that "in the face of lack of access to contraceptives, the change in public attitude, though not necessarily coherent or articulate, must have been resolute" (Glass, 1976). The base of the population pyramid became rapidly narrower as successive generations became progressively smaller. A bulge moved up the population pyramid and by 1981 had disappeared. Successive generations continued to decline in size, if less rapidly, until 1955 when there was a recovery lasting but ten years. In the 1981 population pyramid this can be seen as a bulge at ages 15-24. After 1965 fertility continued its downward trend as can be seen in the 1981 pyramid by the narrowing towards the base. The age structure is no longer youthful; it is very close to that of a stationary population. The percentage of the population under age 15 in 1981 was 20.55; the percentage of those over 65 was 15.08; these values compare with 20.08 and 16.06 for a stationary population.

Before looking forward to the likely or possible changes in the age structure in the future it may be useful to examine in more detail the factors operating to produce these past changes.

FERTILITY CHANGES

It has to be stressed that hitherto the most important factor determining the proportion of older people in the population has not been the improvement in the capacity to survive to old age (though this is by no means negligible) but the changes in the sizes of the generations available *to* survive; i.e. changes in fertility.

If we concern ourselves only with changes in this century, we note that the very sharp decline in the annual number of births in the United Kingdom after 1910 meant that successive generations were progressively smaller so that, as the older generations progressed to pension age, they were replaced by smaller numbers. The *proportion* of younger people in the total population decreased and correspondingly the proportion of

older people increased. It has been shown (Benjamin, 1964) that the change in the proportion aged 65 between 1911 and 1951 can be analysed as follows (Table 2).

Table 2. Change in the proportion of the elderly in the U.K. population, 1911-1951

	%
Aged 65 and over in 1911	5.2
Increases due to:	
abnormal age structure in 1911*	+2.5
fall in fertility alone since 1911	+2.0
fall in mortality alone since 1911	+1.2
Aged 65 and over in 1951	10.9

* The change in the proportion that would have occurred if both fertility and mortality had remained unchanged.

Until 1951, therefore, the total changes in fertility (including the abnormally high fertility prior to 1911) made an overwhelming contribution to the increase in the proportion of the elderly in the population.

Since 1951 the change in the proportion aged 65 and over from 10.9 per cent to 15.1 per cent in 1981 can be apportioned as shown in Table 3.

Table 3. Change in the proportion of the elderly in the U.K. population, 1951-1981

	%
Aged 65 and over in 1951	10.9
Increases due to:	
earlier high fertility and ageing of youthful age structure	+2.7
fall in fertility alone since 1951	+0.2
fall in mortality alone since 1951	+1.6
immigration	-0.3
Age 65 and over in 1981	15.1

It will be seen that mortality improvement now makes a larger contribution to the growth in the proportion of the elderly but that the predominant contribution still comes from the larger generations of births especially prior to 1911 but even after that year which have not been subsequently reproduced.

After 1911, fertility continued to fall. There was a minor recovery in those couples who married just before and during the war of 1939-45 and in the years up to the mid 1960s. Completed family size, either actual or projected

from current fertility trends, has been below population replacement level for marriages in the years since about 1970 and there is no prospect of an early recovery or of any arrest to the progressive ageing of the population structure arising from this cause (Benjamin, 1981).

MORTALITY CHANGES

Meanwhile the ageing of the population structure is being further advanced by mortality improvement. The prospects for mortality improvement were reviewed in 1979 for the purpose of projecting numbers of pensioners (Benjamin & Overton, 1981), and there has not been any significant change to warrant a fresh review.

At that time three mortality projections were made:

Projection No. 1 - Pessimistic Assumptions

For this projection it was assumed that there would be little further improvement in mortality, except at ages below 15 where a somewhat slower rate of improvement than indicated by recent trends was extended into the future. At ages of 15 and above it was assumed that the rate of improvement would quickly taper off to zero. It was not considered that these assumptions would be likely to be realised in the event but that they would provide a lower bound to the widening spectrum of uncertainty in the future.

Projection No. 2 - Medium Projection

This was based on the assumption that the trend of age-specific mortality rates for the United Kingdom in the past two decades had at all ages been approximately such that each year's rate bore a constant ratio to the rate for the previous year, i.e. that for any particular sex/age group the logarithms of the death rates followed a linear trend with time.

These trends were established:

(a) by plotting the death rates on logarithmic paper and fitting a straight line to the plotted points by visual inspection and judgement;

(b) by a least-squares linear fit using an appropriate computer programme applied first to the whole series of death rates and subsequently to more recent years only.

On the basis of a comparison of the result of these three fits which were reasonably close, a judgement was made as to the likely improvement rates which should be applied to the latest available death rates, i.e. those for 1977, to

Table 4. Percentage reduction in mortality rates over a 40-year period

Age	Males GAD 1977	Males Projection No.2 1977	Females GAD 1977	Females Projection No.2 1977
Infant mortality	25	70	25	70
1-4	25	70	25	70
5-9	25	65	25	65
10-14	25	62	25	65
15-19	0	27	0	43
20-24	12	30	14	32
25-29	17	45	25	50
30-34	10		25	
35-39	10	40	10	42
40-44	0		1	
45-49	0	17	0	19
50-54	0		0	
55-59	6	21	3	14
60-64	15		10	
65-69	20	26	14	26
70-74	17		16	
75-79	15	22	17	29
80-84	13		18	
85-89	13	15	19	20

The age-specific mortality rates in 2107 emerging from this projection have been put in the form of an abridged life table as in Table 5.

project rates for the year 2017. These improvement rates, shown as percentage reduction from the 1977 levels, are shown in Table 4. The improvement rates concurrently assumed by the Government Actuary Department (GAD) for national population projections purposes are also shown in Table 4.

The life table shows (given a set of sex-age death rates) the number ℓ_x surviving to age x out of an original ℓ_0 births. The number d_x dying at each age is of course the difference between ℓ_x and ℓ_{x+1}. The average number alive between x and x + 1 is represented by L_x and is approximately $\ell_x + \frac{1}{2}d_x$ (since if the deaths are approximately uniformly distributed over the year of age, the average time survived in the year by the d_x deaths is ½ per year. $\Sigma_0^w L_x = T_x$ where w is the higher possible age in the table is the total years of life lived by ℓ_0 births so that T_x/ℓ_0 is the average life-time or expectation of life and this is represented by $\overset{\circ}{e}$.

Table 5. Abridged life tables based on Projection 2 as at 2017

	Males			Females		
Age x	ℓ_x	$\mathring{e}_x$	T_x	ℓ_x	$\mathring{e}_x$	T_x
0	10,000	74.4	743,774	10,000	80.4	803,758
5	9,948	69.8	693,904	9,958	75.7	753,863
10	9,942	64.8	644,179	9,954	70.7	704,083
15	9,936	59.8	594,484	9,951	65.8	654,320
20	9,906	55.0	544,879	9,942	60.8	604,587
25	9,874	50.2	495,429	9,928	55.9	554,912
30	9,851	45.3	446,116	9,917	51.0	505,299
35	9,824	40.4	396,928	9,901	46.0	455,754
40	9,782	35.6	347,913	9,872	41.2	406,321
45	9,710	30.8	299,183	9,821	36.4	357,088
50	9,515	26.4	251,120	9,695	31.8	308,298
55	9,190	22.2	204,357	9,500	27.4	260,310
60	8,695	18.4	159,644	9,192	23.2	213,580
65	7,921	14.9	118,104	8,725	19.3	168,787
70	6,881	11.8	81,099	8,151	15.5	126,597
75	5,514	9.1	50,111	7,239	12.2	88,122
80	3,857	6.9	26,683	6,030	9.1	54,949
85+	2,266	5.0	11,375	4,401	6.6	28,871

Projection No. 3 - Extreme Assumptions

Dealing first with males, the following extreme assumptions were made as to possible changes over the next forty years:

(a) that deaths from congenital malformations and diseases in early infancy would be reduced to one-third of their present numbers, the lives saved being assumed to die over all ages in proportion to the total deaths of the life table as finally modified in relation to the specified causes;

(b) that as a result of a drastic reduction in the level of cigarette smoking some 90% of all deaths from cancer of the lung and bronchus, and one third of all deaths from ischaemic heart disease prior to age 65 (the proportion currently attributable to smoking) would be saved and proportionately redistributed by age in the finally modified life table.

(c) that as a result of improved therapy and maintenance the whole of the remaining deaths from ischaemic heart disease, from cerebrovascular lesions and other heart and circulatory diseases would be deferred by ten years;

(d) that as a result of environmental improvements including the avoidance of cigarette smoking, all deaths from bronchitis, emphysema and asthma would be prevented and redistributed;

(e) that all deaths from cancers, other than cancer of the lung, would be avoided by the introduction of new therapy, these deaths being redistributed to other causes;

(f) that the risk of death by accident would remain unchanged, some environmental improvements being balanced by the appearance of new hazards (excluding nuclear hazards);

(g) that the small residuum of deaths from tuberculosis and diabetes would remain as at present;

(h) that all deaths in the unspecified cause group would be deferred for ten years, except those in the first year of life which are assumed to be prevented.

The total result of this redistribution of deaths is shown in Table 6 in life-table form.

Table 6. Life table according to Projection 3

	Males			Females		
Age x	ℓ_x	$\mathring{e}_x$	T_x	ℓ_x	$\mathring{e}_x$	T_x
0	10,000	81.3	812,564	10,000	87.1	870,949
5	9,944	76.7	762,704	9,962	82.4	821,044
10	9,935	71.8	713,006	9,955	77.5	791,281
15	9,925	66.8	663,356	9,947	72.5	721,496
20	9,900	62.0	613,793	9,934	67.6	671,793
25	9,870	57.2	564,368	9,924	62.7	622,148
30	9,833	52.4	515,110	9,905	57.8	572,578
35	9,791	49.6	466,050	9,883	52.9	523,105
40	9,728	42.9	417,252	9,845	48.1	473,784
45	9,664	38.2	368,772	9,807	43.3	424,655
50	9,584	33.5	320,652	9,760	38.5	375,737
55	9,486	28.8	272,977	9,710	33.7	327,062
60	9,252	24.4	226,132	9,600	29.0	278,787
65	8,893	20.3	180,769	9,441	24.5	231,184
70	8,365	16.5	137,624	9,194	20.1	184,596
75	7,587	12.9	97,744	8,800	15.9	139,611
80	6,283	10.0	63,069	8,059	12.1	97,463
85	4,673	7.6	35,679	6.881	8.7	60,113
90	2,916	5.7	16,706	5,086	5.9	30,195
95	1,478	3.9	5,721	3,088	3.2	9,760
100	405	2.5	1,013	408	2.5	1,020
105	0	-	0	0	-	0

Modifications to the relative risks of dying from different diseases would normally require the use of multiple decrement methods to

allow for the simultaneous operation of competing risks from the various diseases, but in view of the very arbitrary assumptions of projection 3, refined methods were regarded as being out of place.

A special word must be said about the mortality of females. The current mortality trend for females is a deteriorating one at middle age. Women have been copying the smoking habits of men sufficiently long to begin to incur the penalty of rising death rates from lung cancer and ischaemic heart disease.

So things may be going to get worse before they can get better (this has been implicitly taken into account in projection 2). Nevertheless, having drawn attention to this possibility it was considered possible to apply the same projection 3 assumptions as for men.

The percentage reduction in mortality rates over forty years implied by projection 3 are shown in Table 7 for comparison with Tables 4 and 6.

Table 7. Percentage reduction in mortality rates over a 40-year period: Projection No. 3

Age	Males	Females
0-4	72	75
5-9	48	43
10-14	38	23
15-19	41	27
20-24	38	51
25-29	12	20
30-34	19	33
35-39	15	29
40-44	51	60
45-49	67	71
50-54	77	80
55-59	67	71
60-64	68	73
65-69	70	73
70-74	70	74
75-79	62	70
80-84	61	69
85-89	55	61
90-94	47	54
95-99	25	-5
100-104	66	66

SURVIVAL PROSPECTS

Before considering the population structure consequences of mortality improvement, it is of interest to examine what the likely changes in death rates mean in terms of individual survival. This can be most easily done by referring to the life tables. The deaths at age x, d_x, are the number surviving x years and then dying. So the curve of the function d_x, referred to by actuaries as the "curve of deaths", represents the distribution of lengths of life for the whole generation ℓ_o.

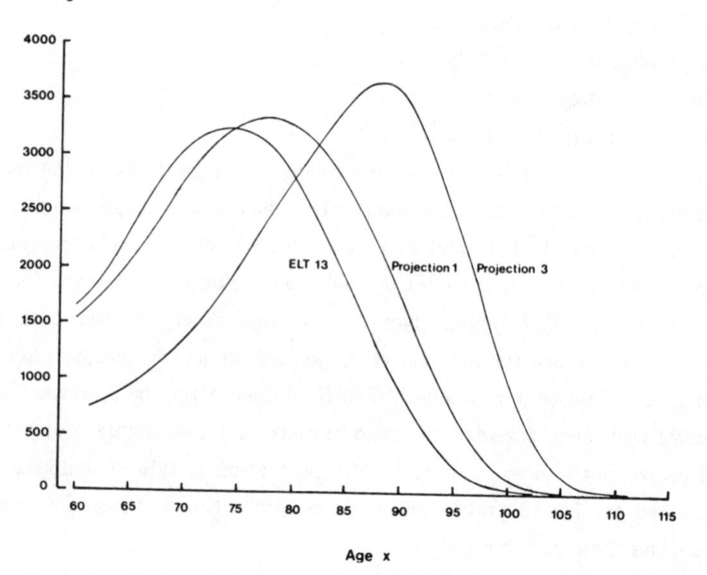

Fig. 2: Curve of deaths in males aged 60 and over, for mortality projections 1 and 3, and national life table ELT 13

Figure 2 shows what these curves look like for mortality projections 2 and 3 and the last official national life table ELT 13. The problem of working with these curves is that they are asymptotic to the age axis and it is difficult to estimate what, for practical purposes, is the age beyond which there are no survivors. It is, therefore, necessary to look for less indeterminate markers that may be used to indicate the extension of the curve of deaths in the direction of advancing age. The following are useful:-

(1) The peak of the curve - the most common age at death.

(2) The point of inflexion in descent, i.e. the second differential equal to zero;

(3) The point of maximum rate of change in gradient before the curve

flattens out at advanced age, i.e. the third differential equal to zero;

(4) The age at which only less than a specified (small) proportion of the original radix of lives is surviving. For this present exercise this proportion has been taken as 10%. (In developing the mortality index of "loss of expected years of life" in 1953 this marker was used as the practical limiting value (Logan & Benjamin, 1953).)

Table 8 shows these markers for a range of mortality tables:

(a) English Life Table No. 11.

(b) English Life Table No. 12.

(c) English Life Table No. 13.

(d) Mortality Projection 2.

(e) Mortality Projection 3.

Looking first at males we see that owing to the past rise in the incidence of ischaemic heart disease and lung cancer the peak of the curve actually moved back slightly between ELT 11 and ELT 13 from 75.7 to 74.5. On projection 2 it moves forward to 76.9 and, not unexpectedly, on projection 3 it leaps forward to 87.7. The point of inflexion has moved only slowly in the past but on projection 2 it will move forward 0.8 years per decade and for projection 3 it will move forward 2.7 years per decade. The final bend after the downward descent of the curve will move forward on the optimistic but reasonable projection 2 to 96.3 and projection 3 takes it to 99.8. On projection 2, 10% of male babies will survive almost to their 90th birthdays and on projection 3, the same proportion survive beyond their 95th birthdays.

For females whose advantage over males in longevity is automatically preserved in the projected tables, the peak of the curve already beyond age 80 will move forward on projection 2 at a slightly slower rate than between ELT 11 and ELT 13 to 86.4 and on projection 3 it moves forward to 95.3. On projection 2 the final bend towards the asymptotic approach to zero is marked at 97.0 and projection 3 takes it to 100.6. On projection 2, 10% of female babies will survive to celebrate their 93rd birthday and on projection 3 the same proportion will celebrate their 97th birthday.

IS THERE A LIMIT TO THE LIFESPAN?

The genetic clock theory of ageing does not demand that every individual be endowed with the same limit of lifespan, or that successive generations should have a constant mean lifespan. It is a fact that no individual has yet been observed credibly to exceed 115 years of age, but then no individual has yet lived

Table 8. 'Curve of deaths' markers

Life table	Central years of application	Peak	Inflexion	Highest rate of change in gradient	10% Surviving
			Males		
ELT 11	1951	75.7	85.8	91.0	84.4
ELT 12	1961	75.4	86.0	91.7	85.1
ELT 13	1971	74.5	86.5	93.3	85.7
Projection 2	2017	76.9	90.2	96.3	88.9
Projection 3	2017	87.7	98.7	99.8	95.6
			Females		
ELT 11	1951	80.3	88.4	93.5	88.0
ELT 12	1961	81.4	89.9	95.3	89.5
ELT 13	1971	82.4	91.3	96.8	90.7
Projection 2	2017	86.4	92.5	97.0	93.4
Projection 3	2017	95.3	99.7	100.6	97.6

in an environment allowing ageing unaffected by disease of external origin. It has been shown in animal studies that there is at least the possibility that suitable adaptation of environment of life style may extend the lifespan. There is a suggestion in Figure 2 that the tail of the curve of deaths *is* pressing towards more advanced ages. This may merely be an approach to a hitherto unattained inextensible limit. Equally it may be possible that there is no limit: that the apparent limit is an accident of history and not a forecast of the future. It may indeed be more profitable to think less about a possible limit and to concern ourselves with the highest age actually attained and the increase in this age which is recorded in the future.

THE CONSEQUENCES FOR POPULATION STRUCTURE

In order to illustrate the effect of the possible trends in fertility and mortality, four population projections have been carried out as detailed below.

Projection	Assumptions	
	Fertility	Mortality
A	TPFR rising fairly quickly to 2.1	Projection 2
B	TPFR rising fairly quickly to 2.1	Projection 3
C	TPFR rising but levelled off at 2.0	Projection 2
D	TPFR rising but levelled off at 2.0	Projection 3

Of the four projections, projection A represents the smallest increase in the proportion of elderly persons and projection D represents the largest.

Table 9 shows the numbers and percentages in various age-sex groups in the actual UK populations for 1981 and as projected to 2021 by population projections A and D. Population pyramids for projections A and D are shown in Figure 3.

It will be seen that on the basis of projection A the proportion in the child population (for this purpose 0-14yr) will fall from its 1981 value of 20.55 per cent to 19.63 per cent and the proportion in the national insurance pension age group (60+ yr for women and 65+ yr for men) will increase from 17.89 per cent in 1981 to 20.04 per cent. On projection D the fall in the child proportion will be greater - to 17.57 per cent; and the rise in the pensioner proportion will be much greater - to 25.44 per cent.

If we look at a more elderly age-group, those 75 and over, we can see that on the medium assumption A the proportion in this age-group will, for men, increase from 1.89 per cent of the total population of both sexes in 1981 to 2.71 per cent in 2021; while, for women, the proportion rises from 3.92 per cent to 4.74 per cent. On the extreme assumption D, the proportion aged 75+ for men rises to 4.73 per cent; and, for women, to 7.11 per cent. In terms of numbers of persons (men and women) aged 75 or over the increase is, on assumption A, from 3266 thousand to 4444 thousand; and on extreme assumption D to 7490 thousand.

ECONOMIC IMPLICATIONS

There are important social and health implications of changes of this scale which are dealt with by other authors in this volume. But there are also

Table 9. Distribution of population at Census 1981 and as projected to 2021 on various assumptions for the United Kingdom.

Sex	Age	1981		2021							
				Assumptions A		Assumptions B		Assumptions C		Assumptions D	
		No.	%	No.	%	No.	%	No.	%	No.	%
Males	0–14	5931	10.54	6028	10.10	6026	9.39	5723	9.74	5720	9.04
	15–24	4510	8.02	3979	6.67	3977	6.19	3829	6.52	3827	6.05
	25–44	7419	13.19	7508	12.58	7491	11.67	7490	12.74	7474	11.81
	45–64	6155	10.94	7447	12.47	7748	12.07	7446	12.67	7747	12.24
	65–74	2267	4.03	2586	4.33	3157	4.92	2586	4.40	3156	4.99
	75+	1062	1.89	1615	2.71	2990	4.66	1615	2.75	2992	4.73
TOTAL		27344	48.61	29163	48.86	31389	48.90	28689	48.82	30916	48.86
Females	0–14	5629	10.01	5689	9.53	5687	8.86	5399	9.19	5398	8.53
	15–24	4347	7.73	3785	6.34	3784	5.89	3645	6.20	3643	5.76
	25–44	7322	13.01	7374	12.35	7365	11.47	7356	12.52	7348	11.61
	45–59	4875	8.67	5919	9.92	6025	9.39	5919	10.07	6023	9.52
	60–64	1583	2.81	1817	3.04	1926	3.00	1817	3.09	1928	3.05
	65–74	2948	5.24	3117	5.22	3521	5.48	3117	5.30	3520	5.56
	75+	2204	3.92	2829	4.74	4497	7.01	2829	4.81	4498	7.11
TOTAL		28908	51.39	30530	51.14	32805	51.10	30082	51.18	32358	51.14
TOTAL (Male + Female)		56252	100.00	59693	100.00	64194	100.00	58771	100.00	63274	100.00

Population of the United Kingdom

Projected to 2021 based on assumption 'A'.

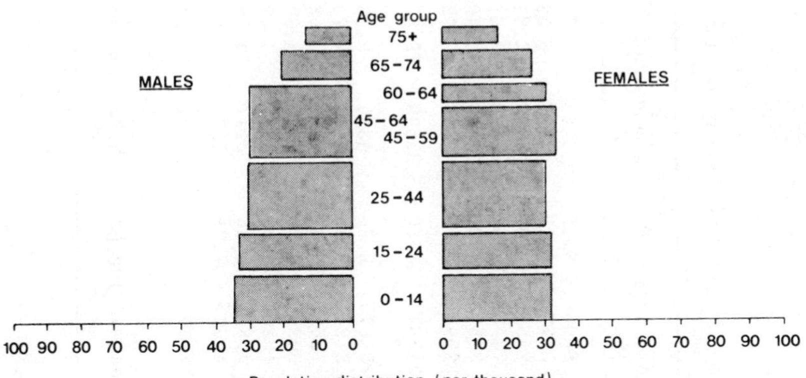

Projected to 2021 based on assumption 'D'.

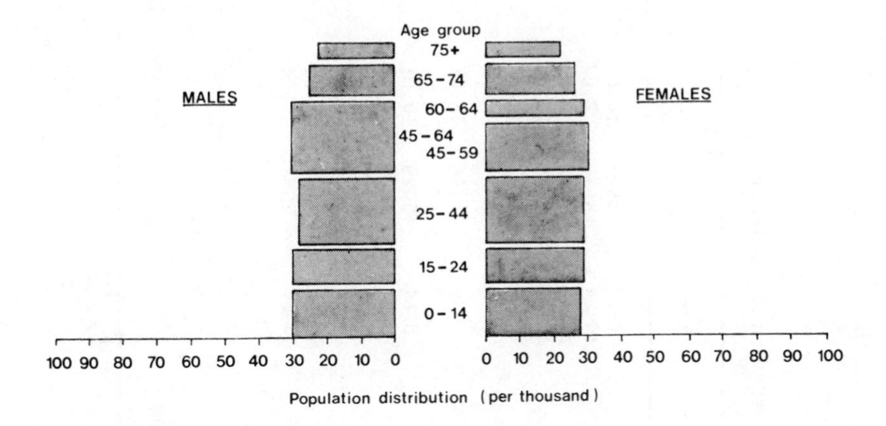

Fig. 3: Projected age distribution of the United Kingdom in the year 2021.

economic implications and since there has been some alarmist talk, for example, about the demographic threat to pensions, perhaps it would be permissible to try to put the record straight.

For illustrative purposes only, we may define three dependent groups (dependent in the sense that although, as in the case of housewives, they may work very hard, other people in paid employment are producing the goods and services they consume).

(1) children under age 15

(2) non-gainfully occupied women

(3) persons who have attained the current minimum retirement ages (for National Insurance purposes) of 65 (men) or 60 (women).

The balance may be regarded as indicative of the supporting sections of the population, though it should be borne in mind that, at the present time and possibly for some time in the future, more than one-tenth of this balance are themselves dependents as members of the unemployed. We refer to this reservation again later.

Changes in the number and proportions in these groups are shown in Table 10.

It might be argued immediately that

(a) these ratios are not true measures of dependency; and

(b) the ratios are so incomparable in real economic terms that they should not be added together to produce the overall index of column (10) (Table 10).

As to (a) it should be emphasised that these ratios are indices and *not* absolute measures. The relative changes over time are important; absolute values are not. As to (b) it is agreed that a straight addition of the indices may not be supportable. On the other hand, if a weighted combination were required what should be used? If one considers all the goods and services (especially services) consumed by the various groups it becomes very difficult to argue that they are substantially incomparable.

With all its defects a straight addition seems as justifiable as a weighted addition. Again, no *absolute* meaning should be attached to the index values in column (10); we are only concerned with relative changes over time.

Several interesting features emerge from Table 10. The first is that the total pressure of national dependency (column 10) declined between 1901 and 1981, and only then began to rise. By 1971 however, the growth in the total pensioner dependency had been matched by the

Table 10. The economic pressure of the dependent population in the United Kingdom (in thousands)

Projection	Year	Home population all ages	Children under 15	Non-gainfully occupied females 15–59	Pensionable class i.e. males 65+ females 60+	Remainder	Ratio to Remainder of Col. (3)	Col. (4)	Col. (5)	Cols. (3+4+5)	Proportion of unemployed to economically active population
	(1)	(2)	(3)	(4)	(5)	(6)	(7)	(8)	(9)	(10)	(11)
	1901	38,237	12,421	7,717	2,386	15,713	.79	.49	.15	1.43	
	1921	44,027	12,304	9,560	3,501	18,662	.66	.51	.19	1.36	0.093(b)
	1931	46,038	11,174	9,819	4,421	20,624	.54	.48	.21	1.23	0.110(b)
	1951	50,225	11,325	9,211	6,827	22,862	.50	.40	.30	1.20	0.009(b)
	1961	52,709	12,336	8,629	7,733	24,011	.51	.36	.32	1.19	0.012(b)
	1971	55,515	13,387	7,325(a)	9,015	25,788	.52	.28	.35	1.15	0.035(b)
	1981	56,252	11,560	8,272(a)	10,064	26,356	.44	.31	.38	1.13	0.114(b)
A	1991	56,948	11,140	8,455(a)	10,479	26,874	.41	.31	.39	1.11	
[Medium]	2001	58,087	12,277	8,538(a)	10,259	27,013	.45	.32	.38	1.15	
	2011	58,639	11,490	8,615(a)	10,902	27,632	.42	.31	.39	1.12	
	2021	59,693	11,717	8,539(a)	11,964	27,473	.43	.31	.44	1.18	
D	1991	57,649	11,140	8,468(a)	11,085	26,956	.41	.31	.41	1.13	
[Extreme]	2001	59,777	12,117	8,566(a)	11,902	27,192	.45	.32	.44	1.21	
	2011	61,252	11,019	8,643(a)	13,719	27,871	.40	.31	.49	1.20	
	2021	63,274	11,118	8,507(a)	16,094	27,555	.40	.31	.58	1.29	

(a) Estimated as 50% of age group
(b) Census figures

increased employment of women (mainly married women) and the overall index had fallen below the level of 1931. (This is a quite remarkable development in the British economy.) In 1971 the total dependency ratio was below that for 1931 and below that for 1951. In 1981 the total dependency ratio was slightly lower than in 1971.

Over the next 40 years on the moderate assumptions of projection A, it appears that child dependency will decline slightly and pensioner dependency will increase by about one-sixth, while the dependency of non-earning women will remain unchanged. This will leave the overall dependency index higher than in 1981 but not greatly so; and only a little higher than in 1971. It is important to note, too, that (on projection A) pensioner dependency does not rise until after 2011. There is therefore still a decade or two in which to plan ahead for the additional economic strain. Even on the extreme assumptions of projection D, though pensioner dependency will be much higher than it has ever been, the overall dependency index will not be much larger than it was in 1951 (when the UK economy was recovering from war and when, unlike now, manpower was in short supply).

The rise in pensioner dependency, too, is not substantial until after the end of the century. Moreover, we have to emphasise that the extreme assumptions, which form the basis of projection D, really are extreme. The majority of observers regard the outcome as likely to be closer to A than to D.

In summary, therefore, a decline in the birth rate and some mortality improvement (particularly a reduction in child mortality) have already been encompassed once this century without serious economic strain, and claims that further demographic changes will be economically intolerable do not seem to be founded on past experience, or on the facts as to likely changes in population structure.

Moreover, it has to be stressed immediately that these dependency ratios for 1981 and in the future are not comparable with earlier values for the reason that technological improvements have greatly reduced the number of workers needed to produce the goods and services required by pensioners - if this were not so we should not be seeing such a serious rise in unemployment. And even if the same number of workers were required as formerly, then they are there - some 3-4 millions of them - waiting for work.

THE DEPENDENCY OF THE UNEMPLOYED

The tragedy of it all is that there is now, in addition to the elements of dependency so far considered, another element which has become a major

concern - the dependency of the unemployed. Column 11 of Table 10 shows the ratio of the unemployed to the economically active population. This is after all a part of total dependency which cannot be blamed on demography or pension rights.

This paper is specifically about numbers. One is well aware that there is more to the total problem of caring for the elderly than their numbers. It is necessary to answer those, mainly politicians, who have shown an inclination to use the rise in the proportion of the elderly to suggest that pension benefits in the future cannot be supported and should be substantially reduced. Alternatively there have been suggestions (improvident in view of the world population pressure) that financial incentives to produce larger numbers of children should be introduced. Given intelligent advance planning there is no need for such over-reaction.

THE AGEING OF INDIVIDUALS

There is another demographic aspect of the social problem of ageing that is not discussed by those who are worried by mere numbers of old people relative to the young and which is worth discussing further. This is the problem of ageing for the individual.

There are difficulties in extending the working life even if employment were available. Although mortality has been considerably reduced it is not easy to interpret this increased longevity in terms of prolonged activity, i.e. in terms of the quality of survival. Postponement of death in an elderly person by improved medical care does not necessarily imply the arrest of the relentless process of degeneration. We do not know whether the decline in mortality is an extension of healthy life or merely the postponement of death in sick people by maintenance therapy.

Statistics supplied by the Government Actuary (1984) (personal communication) show that the proportion of the insured male population who had been sick for more than three months at mid-1982 rose from 3.3 per cent at ages 45-49 to 4.5 per cent at ages 50-54. The proportion begins to rise rapidly after age 60, and even before 65 more than one-seventh of insured men are in poor enough shape to have been on sickness benefit for more than three months.

On the other hand, surveys of the aged population have indicated that infirmity is more quickly developed and more passively accepted in conditions of stagnation and boredom and is more effectively resisted and prevented by interest and occupation. As long ago as in 1947 the Committee on the Problems

of Ageing and the Care of Old People reported that they had been "impressed by the views expressed to them of the higher therapeutic value of occupation and employment in delaying the effects of ageing and they feel that it is in the interests of those who are elderly but not old to be able to continue in employment as long as they wish to do so". Many others have reported in the same vein.

THE EXTENT OF EMPLOYMENT

There is indeed no reason why there should be passive acceptance of invalidity either by the individual or by the community. It may be that the proportion of workers who pursue employment beyond normal retiring age is not as large as it could be. Estimates of the proportion of males in employment in the older age groups, derived from the 1981 Census, reflect the influence of pensions in facilitating retirement at age 65. At the 1921 Census, when there was some economic depression but no widespread system of pensions to influence retirement from work, the proportions employed beyond age 65 were considerably higher. Whether this is evidence of the sustaining influence of occupation, or whether it merely represents the less definite transition from full-time working through stages of unfitness to retirement which existed before pensions provision became widespread, is difficult to judge.

The real problem which the statistics do not at present measure is the extent to which the economy of the country fails to match the limited and specialised aptitudes of older workers to equally limited and specialised occupations so that they make the maximum possible contribution to the national product within their physical capacity and enjoy doing it. It ought, for example, to be possible to find work for them to do in the social services to relieve the pressures of dependency.

CONCLUSION

To repeat what has been said earlier, this paper is about increasing numbers and proportions of old people. It does not touch on the social and health services problems of a growing elderly population. These important problems are to be dealt with by other speakers. But one more number may be left with you. Assuming that fertility does not fall any further by the year 2061 most of the effects of the decline of births will have worked their way through the population. The effects of mortality decline will still be continuing but these changes have been less sharp than the fall in fertility and are likely to remain gradual giving more time for them to be accommodated.

REFERENCES

Benjamin, B. (1964). Demographic aspects of ageing. Journal of the Institute of Actuaries, **90,** 213-253.

Benjamin, B. (1981). Recent and prospective fertility trends in Great Britain, In: D. F. Roberts & R. Chester (eds.), Changing Patterns of Conception and Fertility. Report of the Proceedings of the 19th Eugenics Society Symposium. London: Academic Press.

Benjamin, B. & Overton, E. (1981). Prospects for mortality decline in England and Wales. Population Trends, **23,** 22-28.

Committee on the Problems of Ageing and the Care of Old People (1974). London: Nuffield Foundation.

Glass, D.V. (1976). Recent and prospective trends in fertility in developed countries. Philosophical Transactions of the Royal Society. **274** B, 9.

Logan, W.P.D. & Benjamin, B. (1953). Loss of expected years of life. Monthly Bulletin of the Ministry of Health and Public Health Laboratory Services, Issue No. 12, 244-248.

WHERE DO OLD PEOPLE COME FROM? AN EVALUATION OF AMERICAN POPULATION PROJECTIONS

J. S. MacDONALD

King's College, University of London,
London, U.K.

The recent upsurge of interest in geriatrics and gerontology throughout the Western world, in the Soviet bloc countries and also in Japan, is not based only on the larger presence of the elderly and their better articulated demands. It is also based on the expectation that this proportion of elderly will always be with us and, indeed, will continue to increase. Let us see whether the expectation that geriatrics and gerontology are a growth industry is well founded. The population projections for the United States, Current Population Reports Series P25 No. 952 issued by the Census Bureau in June 1984, are a good case in point. The United States' demographic processes are similar to many other urban-industrial countries'; and it has great cultural and technological influence on other urban-industrial countries in the setting of demographic trends. Moreover, the new American projections are extraordinarily elaborate and relatively clearly explained.

Let us examine the fundamentals of the United States projections in relation to the processes tending to make for faster or slower demographic ageing. In doing so, we can find some of the answers to the general question: Where do old people come from? and more pointedly: Why do they come up the population pyramid in larger or smaller numbers in different periods? Figure 1 shows the successive re-shaping of the American population pyramid as it has grown progressively older; and also shows its future age composition according to the medium population projection which foresees further increases in the proportion of people aged 65 and over.

It is customary to accept population projections rather blindly or to dismiss them lightly. Certainly some scepticism can be justified: the majority of professional demographers intentionally avoid making population projections and many are outspoken in their criticism; and, typically a few years after a population projection is issued, the actual course of events fails to conform and not infrequently contradicts it flatly. Yet, geriatricians and gerontologists are

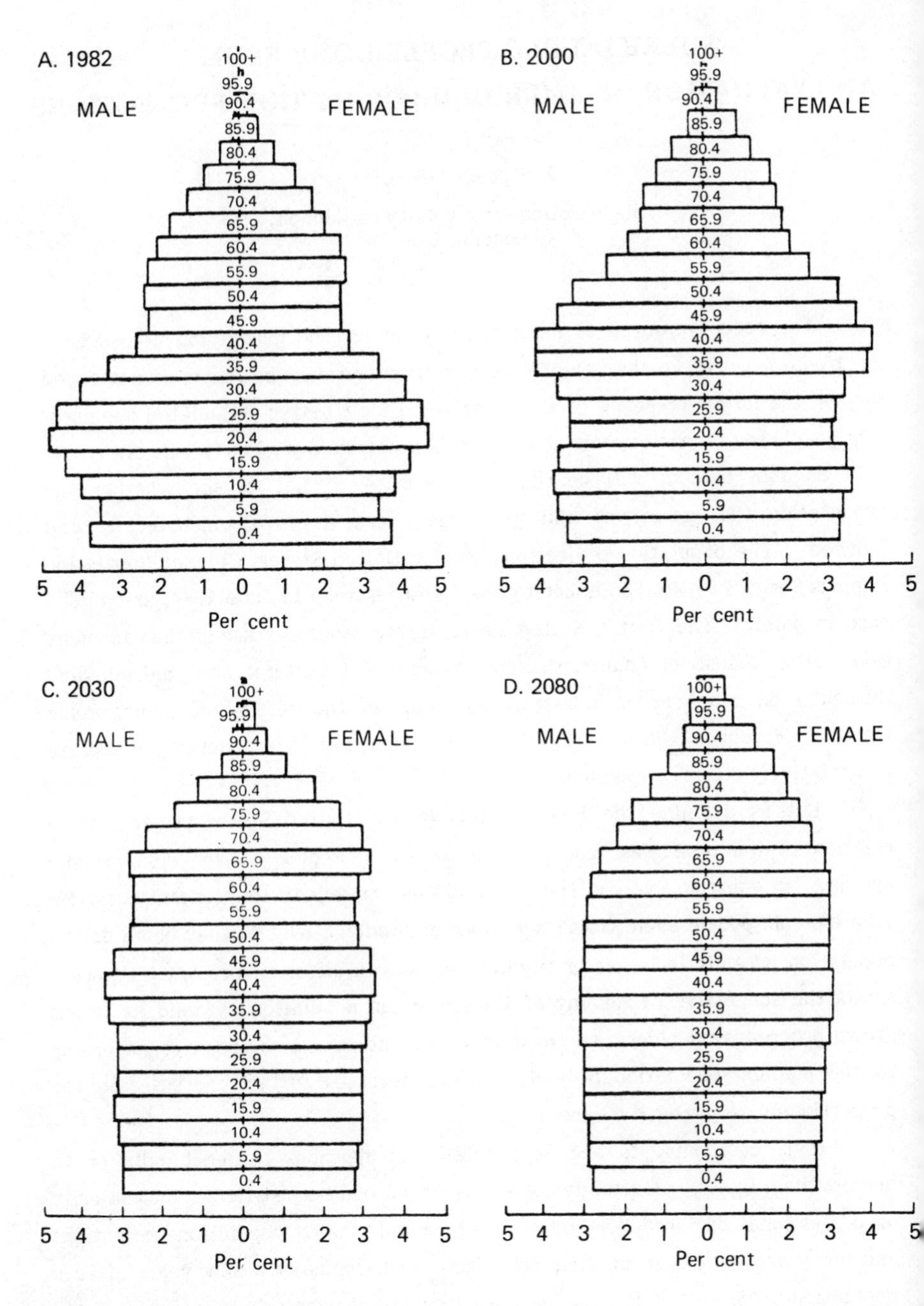

Source: US Bureau of the Census.

Figure 1. Population pyramid: percent distribution by age and sex.

Table 1. Projections leading to highest, lowest or middle population growth: percent aged 65 and over

Year	Lowest	Middle	Highest
1985	12.0	12.0	10.9
1990	12.8	12.7	12.6
1995	13.2	13.1	12.9
2000	13.1	13.0	12.9
2005	13.2	13.1	13.0
2010	14.0	13.8	13.6
2015	15.8	15.4	14.9

Lowest assumptions: low fertility, low immigration, high mortality. Middle assumptions: all medium. Highest assumptions: high fertility, high immigration, low mortality.

Source: US Bureau of the Census, 1984.

Table 2. Proportion of elderly, 1900-1982

	% aged 65+
1900	4.1
1910	4.3
1920	4.7
1930	5.4
1940	6.8
1945	7.2
1952	8.1
1955	8.8
1960	9.2
1965	9.5
1970	9.8
1975	10.5
1980	11.3
1982	11.6

Source: US Bureau of the Census (various series)

generally unquestioning in their acceptance of population projections, at least when they point to larger and larger proportions of the elderly in the future.

Certainly the explanation and evaluation of population projections make a formidable task because of the immensity of the data handling. It is for good reason that only governments and international agencies normally make population projections rather than smaller institutions. Perhaps the wider availability

of less forbidding computers will soon diffuse the art of population projections, but, at the moment, it remains an esoteric pursuit to be viewed by most of us simply from the outside. Therefore, this paper is, incidentally, intended to serve as a readers' guide to the latest United States projections of the elderly population.

These projections are based on the total population and on the fertility and immigration statistics in mid-1982 and on the death statistics up to October 1982. They are a complete revision of the 1977 population projections, rethinking the most recent trends of births, deaths and net immigration and also incorporating the latest generalisations and techniques from professional demography. Although 30 projections were calculated and are briefly referred to in the report, the official presentation emphasises the proportions of elderly people foreseen by the three types of projections in Table 1. The high projection is high in the sense that it leads to the fastest and greatest population growth and, in the same sense, the low projection points towards slower, smaller population growth. The Bureau report inclines towards the middle range population projection, which is presented more often than the two extremes, which are thereby lent less credence. However, these three types of projections are not to be examined here from the standpoint of their relative contributions to total population growth, but in terms of their relative contributions to the proportion of elderly people in the future.

Table 2 shows that the demographic ageing of America's population goes back to 1900 and, in fact, the proportion of the population aged 65 and over had been going up persistently since 1870. The medium, high and low population projections all indicate continued demographic ageing well into the next century. However, none of the three anticipated up-trends is as straightforward as the past: the high population projection leads to a levelling off of ageing towards the late 1990's; the medium and low population projections deviate a little further, producing a slight drop in the proportion of elderly by the year 2000. Despite the faltering of the trend at the turn of the century, it is clear that there is a master trend at work since the last century. Here we must ask: Does the considerable variety of options for change allowed for by the three officially preferred projections in fact give room for all potential demographic ageing processes? Why does the master trend falter for the first time about the end of the century?

All the official population projections are calculated to the year 2080, but the discussion will be limited here to the period up to 2015. It seems unrealistic

Table 3. Components of population projections, 1982–2080

		Initial Level	Population Assumptions								
			Low			Medium			High		
		1982	2005	2015	2080	2005	2015	2080	2005	2015	2080
Total fertility rate per 1000 women, 15–49		1831	1662	1641	1600	1961	1949	1900	2302	2350	2300
Life expectancy at birth (in yrs)	M	70.6	75.3	76.0	80.8	73.6	74.0	76.7	72.0	72.2	73.5
	F	78.1	83.2	84.2	91.0	81.2	81.7	85.2	79.5	79.8	81.2
Net immigration per year (1000s)		258	250	250	250	450	450	450	750	750	750

Source: US Bureau of the Census, 1984.

to go further into the future: in the year 2015, the last of the women entering the child-bearing years in the middle of this decade will have approached the end of their reproductive capacity at age 45. As the discussion will point out, the fluctuations of fertility are very hard to explain and predict, to judge by the last fifty years of American demographic development. Therefore it is wise and practical to limit the future extent of projections to no more than a span of one cohort of women passing through its child-bearing years. Indeed, the Census Bureau works out the interplay of the 3 components - fertility, immigration and mortality - in full detail only up to the year 2005, and thereafter calculates relatively simple trajectories which, by the time they pass 2015, enter rather blindly into combinations unrelated to any foreseeable contextual determinants. While population projections to the year 2015 and the components put into them are themselves somewhat questionable at each step, certainly developments beyond that year should be left to science fiction.

As Table 3 shows, the medium population projection, which is the most preferred officially, is calculated from a total fertility rate passing from 1,831 births per 1,000 women in 1982 to 1,949 in 2015, a net immigration balance of 450,000 per year throughout, and life expectancy passing from 70.6 years in 1982 to 74.0 in 2015. The medium assumptions about future fertility and immigration are very much extrapolations of the immediate potential of their recent levels and trends, without much re-thinking to estimate the changing context of fertility or the errors in the net immigration statistics. Setting the sights of the fertility component of the medium population projection so that it will reach 1.9 life time births per woman is simply spelling out the extremely slight lift in fertility since the historic low of the mid-1970's. The medium net immigration balance of 450,000 per year is kept at the same level throughout the medium projections, simply repeating the average recorded in 1973-1982.

By contrast, the medium mortality assumption moves quickly away from simply extrapolating: it does not plot an extension of the extraordinarily rapid reduction of mortality between 1968 and 1982 which was at the rate of 0.43% of initial life expectancy per calendar year. Instead, the pace of mortality decline until 2005 is set at 0.18% of life expectancy per year, that is only at about 40% of the pace of the actual decline in the previous 15 years. After 2005, the medium mortality assumption is set even further above the recent trend line, namely at the rate of 0.6% of life expectancy per year. We shall return to the rationale for these great departures from recent mortality experience. In other words, the 1982 life expectancy of 70.6 years for males and 78.1 for females is

projected to rise to 75.3 and 83.2 years respectively by 2005. Then, until 2080, there would be the very slow gains of only 6.1 years for males and 7.1 for females.

Why do these projected middle range trends lead to an increase of the elderly population from 11.6% in 1982 to 13.1% in the year 2005 and 15.4% in the year 2015?

IMMIGRATION LEVELS AND TRENDS INTO THE FUTURE

Let us first examine the immigration assumption more closely because it is the most problematic of the three components. The choice of an assumption about net immigration is less well supported by the statistical system than the selection of trend lines for fertility or mortality. While birth and death registration in the United States has been quite complete since 1940 and highly reliable even earlier, emigration statistics have not been thoroughly compiled and have been scarcely analysed. Immigration statistics miss the apparently large margin of illegal immigration. Moreover immigration, analytically, is less autonomous and less explicable than fertility or mortality because it is more immediately and exclusively determined by the context of economic, political and social changes. The determinant context of immigration not only includes the interplay of economic, political and social factors within the United States, but also in the sending countries. Faced with the daunting task of making sense of this most elusive component, the Census Bureau medium population projection simply averages recorded immigration into the future, without paying any attention to either illegal immigration or unrecorded emigration.

The choice of medium immigration level as the trend into the future is political in a sense. Immigation is a very sensitive issue: there has been a vigorous debate, often extremely heated, in recent years about the real extent of illegal immigration. The medium immigration projection avoids this major element of America's total immigration, as if it could be closed off. Indeed there have been moves in Congress to minimise this inflow; and, conceivably, this campaign could have been passed into law and into fairly effective administrative procedures. This has not yet occurred despite repeated raising of the issue. For the time being, it seems realistic not to overlook the high immigration assumption which does allow for illegal immigration.

What difference does it make to the percentage of the elderly in the medium population projection if the high annual net immigration figure of 750,000 is put instead of 450,000? Because total net immigration is not properly

Table 4. Age composition of net immigration as a percentage of the net immigration balance

Ages	Assumption Low	Medium	High
0-17	36.5	30.6	31.4
18-24	15.1	17.1	17.0
25-29	21.7	16.6	18.2
30-34	11.8	10.9	11.2
35-44	9.4	11.6	10.9
45-54	5.4	6.7	6.1
55+	-0.1	6.2	5.3
	100.0	100.0	100.0

Source: U.S. Bureau of the Census, 1984.

recorded, its age composition is no better known than its volume. Despite these gaping holes in the data-base, sensible estimates, albeit tentative, can be made of the effect of net immigration's age composition on the oldness of the total population pyramid. While commonplaces are suspect in demography, it really does look as if commonsense is correct in deducing that net immigration rejuvenates a population pyramid in the short run and has little or no effect on ageing in the long run. For recorded immigration rarely occurs beyond age 35 and is concentrated between 18 and 30.

For example, in the case of an unsettled labour immigration stream, most arrivals are aged 18-29, consisting mainly of bread-winners. Some of their dependants, including children, may not follow them for years if ever. Another example: in the case of a more settled immigration stream, largely comprising family groups, arrivals are more spread out between 0 and 34, with a tail of older dependants re-joining them later. It seems a fair guess that illegal immigration, like recorded immigration, is heavily concentrated between 18 and 35. Deduction suggests that emigration pares down the higher ages of the population pyramid by exporting ageing immigrants and also that the age composition of native-born emigrants is considerably older than the total population pyramid. To the extent that the Census Bureau underestimates emigration, it underestimates the youthfulness of the net immigration balance.

As Table 4 demonstrates, the Census Bureau estimates that net immigration will comprise between 45 and 49% in the younger working ages, 18-34, depending on the choice of high, medium or low net immigration assumptions. Minors aged 0-17 would range between 31 and 37%. So the proportions aged 35 and over range up to 25%, according to the medium and high immigration assumptions, and may be only 15%, according to the low immigration assumption.

Table 5. Net immigration and percent aged 65 and over: high, medium and low immigration assumptions

	Low immigration Medium fertility Medium mortality	Medium immigration Medium fertility Medium mortality	High immigration Medium fertility Medium mortality
1985	12.0	12.0	12.0
1990	12.7	12.7	12.6
1995	13.1	13.1	12.9
2000	13.1	13.0	12.8
2005	13.2	13.1	12.9
2010	14.0	13.8	13.5
2015	15.6	15.4	14.9

Source: U.S. Bureau of the Census, 1984.

Substituting the high for the medium immigration assumption, but leaving fertility and mortality at the medium level, reduces the proportion of the elderly in 1995 and again in 2005 from 13.1% to 12.9%. In 2015, the elderly are reduced from 15.4% to 14.9% as is shown by Table 5. In other words, a larger net immigration input, including a crude but more realistic estimate for illegal immigration, makes the population younger for the next three decades.

The ageing process could be increased by immigration only after a long delay that is beyond 2015. In any case, the long run effect obtains only if immigration takes the form of a wave climbing rapidly above a low point or falling away sharply into a trough. Even so, the effect of such a steep-sided wave is substantially smoothed down by the child-bearing of immigrants, especially in the case of settled family immigration which has been the typical form of immigration into the United States. In fact, postwar net immigration in the United States has not taken this pronounced shape for a period long enough to be of consequence as far as the records show. The delayed-action effects of a sharp-angled wave of immigration take so long to work their way up the population pyramid that intervening processes usually muffle the repercussions. For example, the population pyramids of the postwar decades exhibit no clear effect of the sudden halting of mass immigration in 1914-1918. Another example: the peak refugee inflows from Hungary in 1957, from Vietnam in 1974 and from Cuba in 1980 were neither high enough nor wide enough nor age-selective enough to override the dominance of fertility's effects on the shape of the population pyramid in the short or the long term.

It is notable that the inflow of 750,000 immigrants net each year posited by the high immigration assumption does not alter the levelling off of the

proportion of elderly between 1995 and 2005 which appeared in the all-medium population projection. Thus, recognition of the possibility of massive continuing immigration - or rather the recognition of the actuality of massive continuing illegal immigration - has very little effect on the trend-line of the proportion of the elderly up or down. That is, high immigration does no more to *reverse* demographic ageing than medium immigration, but high immigration does *decelerate* ageing relative to medium immigration.

The historical record shows that a fall in fertility and a fall in immigration together can have a considerable delayed effect on demographic ageing, but there is no simple instance of immigration alone having such a marked effect on the ageing of the United States population. Indeed, as we have seen, the shrinking of the proportion of elderly towards the year 2000 is due to the Great Depression of the early 1930's when net immigration became negative and fertility fell below replacement. A clearer case occurred after the Second World War when the survivors of the great transatlantic immigration which had peaked between the 1890's and 1914 reached retirement age. This great immigration was stopped by the First World War, resumed at a reduced level in 1919-1921, but was kept down by immigration restrictions until the Great Depression and then the Second World War. However, the acceleration of demographic ageing in the United States between 1945 and 1950, when the percentage rose 0.9% from 7.2 was not simply due to the wave of immigration passing age 65, but rather to a combination of this immigration effect with the low levels of fertility between 1929 and 1936 which had depleted the age bracket 0-19, lightening the young end of the population pyramid so that the balance slipped towards the elderly faster than the already long-standing demographic ageing process.

Even in the peak years of relatively well-recorded immigration after the Second World War, the official net immigration balance comprising refugees plus other immigrants has been well below one-tenth of the births until the late 1960's when, in some years, it surged to 12.5%. In the 1970's the official figures are especially questionable. It was then that illegal immigration loomed large. Nevertheless, if total immigration, recorded and unrecorded, has averaged 750,000 since 1970, it still did not exceed 25% of births in any one year. Since fertility's impact on the population pyramid is concentrated on age 0 and then passed on to early childhood, while immigration's impact is much more diffuse over ages 18-29 and, to a lesser extent 0-34, it is quite clear that fertility has been determining demographic ageing much more than net immigration even in

recent years of rising immigration.

What difference would it make if, instead, immigration were reduced to, say, 250,000 in accordance with the Census Bureau's low immigration assumption? This could conceivably occur through narrow administrative restrictions, like those applied until 1965, combined with economic depression. We have seen in Table 4 that this low net intake each year includes negative net immigration above age 55, and proportionately far fewer arrivals above age 35 than in the medium or high immigration assumptions. In Table 5, there is a definite spread between the proportions of the elderly produced by the low immigration projection and the high immigation projection, for example, 13.2 versus 12.9% in 2005 and 15.6 versus 14.9% in 2015. That is, the combining of low net immigration with medium fertility and medium mortality has the effect of raising the proportion of the elderly: the absence of high or medium immigration tips the population pyramid towards greater ageing; this effect is quite substantial by the year 2000.

FERTILITY LEVELS AND TRENDS INTO THE FUTURE

The wide fluctuations of fertility in the United States (and other North Atlantic countries) since the Great Depression make projections of future fertility much more difficult than in the decades before the 1930's when fertility followed a remarkably simple trajectory downwards. Fortunately for the demographer, the American statistics of fertility are excellent, unlike those for immigration or emigration. Moreover, fertility is not subject to the international contextual factors - economic, political or social - which are such important determinants of net immigration. There are numerous thorough analytic studies of the economic, political and social conditions of high and low fertility in the United States especially since 1960, while the literature on immigration is poor. Therefore the fertility projections are less problematic than the net immigration projections, although they are heavy with debatable generalisations.

The high fertility assumption in Table 6 reduces the trend towards an older population which we have already seen in the medium fertility projection. Moreover this high fertility variant does reverse the overriding surge towards great demographic ageing for a longer period than the high immigration variant at the turn of this century. Is this not a contradiction of the generalisation that fertility is the demographic component with the strongest effects on the demographic ageing process? The answer is that we must distinguish between

Table 6. Fertility and percent aged 65 and over

	Low fertility Medium immigration Medium mortality	Medium fertility Medium immigration Medium mortality	High fertility Medium immigration Medium mortality
1985	12.0	12.0	12.0
1990	12.8	12.7	12.6
1995	13.3	13.1	12.9
2000	13.4	13.0	12.8
2005	13.6	13.1	12.8
2010	14.5	13.8	13.3
2015	16.3	15.4	14.5

Source: US Bureau of the Census, 1984.

the short-term effects of fertility and the long-term effects of fertility. In the short term, higher fertility tips the balance of the population pyramid towards the younger age brackets because, obviously, fertility has its impact first on age 0. Immigration, by contrast, is most weighty towards the middle of the population pyramid, thereby having a more diffuse effect than fertility, which has a see-saw tipping effect on the age structure.

Historically, it was the down-trend of fertility since the early nineteenth century, reaching a trough in the 1930's, that produced the demographic ageing of America's population (and indeed the demographic ageing of all North Atlantic populations) up to the post-war baby boom. Tracing the historic trend lines of the three demographic processes shows that it could not have been either immigration or mortality which caused the remarkably steady increase in the proportion of elderly since the middle of the last century. The historic mortality downtrend since the middle of last century has continued up to the present with only a very slight pause between the mid-1950's and the mid-1960's. This great consistency of the mortality downtrend might suggest that it was mortality which, being highly correlated with the equally consistent up-trend of demographic ageing, was the cause of the latter. To some, it may still seem strange, even absurd, that the proportion of elderly people should depend on births rather than deaths. Indeed, until the 1950's, it was commonplace in professional demography, as well as among the lay public, that the increasing proportion of elderly people was due to increasing life expectancy. Then demographers demonstrated a paradox: reduction of mortality, history showed, increased the survival of potential mothers and thereby increased births more than it increased the proportion of the elderly. It will be seen in the more thorough discussion of mortality later in this paper that this generalisation must be substantially

Table 7. Differences in percent of elderly between high and low assumptions with the other two assumptions held at middle level

	Component		
	Fertility	Net immigration	Mortality
1995	0.4	0.2	0.2
2005	0.8	0.3	0.6
2015	1.8	0.7	0.9

Source: U.S. Bureau of the Census, 1984.

amended for the mortality effects on ageing over the last 15 years and also for the mortality effects on ageing in the projections for the future. But the important point here is that fertility's short and long term effects had, since the last century, been at work on the building up of the 1982 population pyramid which is the base line for the latest population projections.

Table 7 isolates the consequences for demographic ageing stemming from the high, medium and low fertility assumptions. These three trends of fertility create a wider spread between estimates of the proportion of elderly in future than do the difference among the high, medium and low assumptions for immigration or mortality. By holding migration and mortality at the middle level, Table 6 demonstrates how much the high fertility assumption, with its uptrend from the total fertility rate of 1,831 births per thousand women in 1982 to a total fertility rate of 2,302 in 2005 and 2,350 in 2015, could slow demographic ageing. While the low fertility projection leads to 13.6% elderly in 2005 and 16.3% in 2015, the high fertility projection results in 12.8% elderly in 2005 and 14.5% by 2015. Thus the spread of 0.8% between high and low fertility effects in 2005 is wider than the spread between high and low immigration effects or between high and low mortality effects calculated by the same method of combining one high assumption with two medium assumptions. By 2015, fertility clearly makes much more difference than immigration or mortality.

The high fertility assumption is not very high compared to the peaks of fertility in the postwar period. As Figure 2 implies, when the women born in the depths of the Great Depression were in the vanguard of the postwar baby boom, they had the highest completed family size of any this century, 3.1 children, despite the fact that they were daughters of the cohort with the very lowest completed family size in American history (and in North Atlantic history) up to that time. This sharp reversal of the long established fertility downtrend in

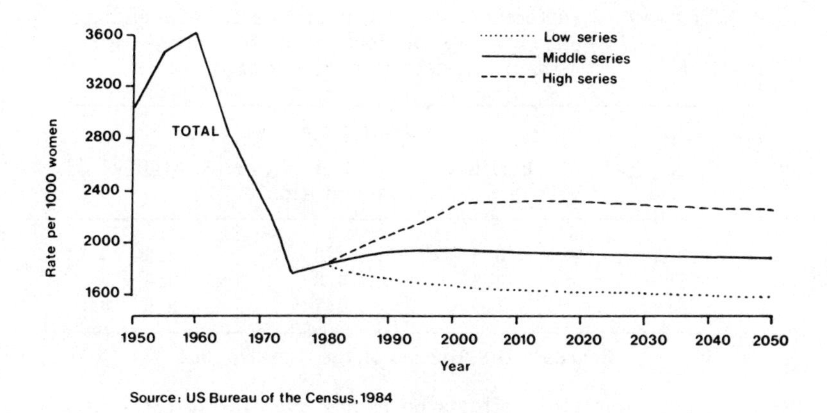

Figure 2. Estimates and projections of annual total fertility rate, 1950–2050. Approximate potential completed family size of cohorts.
Source: US Bureau of the Census, 1984.

America (and in the North Atlantic countries) is a sharp reminder of the volatility of fertility over the last 50 years. It is a reminder especially that the currently plausible range between high and low fertility assumptions may quite conceivably be left behind by a wider swing to the levels of the postwar baby boom before 2015. If completed family size were to swing back beyond the high fertility assumption of 2.3 children and approach an average of 3 children or more, demographic ageing would slow down even further than under the high fertility assumption.

Why did the Census Bureau decide not to use a higher *high* fertility assumption? Why did they not decide to use a high *medium* fertility assumption? A higher high fertility assumption or a higher medium fertility assumption would lead to smaller proportions of the elderly in the period under consideration, that is until 2015. It is taken for granted in many North Atlantic countries' official projections that fertility will return to the replacement level of 2.1 children on the principle that a nation should not lose population. In previous projections, the Census Bureau followed this practice in its medium fertility assumption. The present all-medium population projection is based on an assumption of a completed family size of 1.9 children ultimately. In this historic decision, it has now scaled down the ultimate completed family size in each fertility assumption on the following grounds: (a) Surveys show that the number of births expected by women in the early childbearing years remains below 2.1 with no rising trend apparent. In any case, the number of births

expected by single women and childless women have repeatedly proved later to be higher than their actual completed family size, as previous surveys of family building intentions demonstrate. (b) Women's position in the household, in the education system and in the labour market has been changing in directions strongly correlated with smaller completed family size: (i) female labour force participation has continued to rise; (ii) the age of women at their first marriage has continued to rise; (iii) more married women are becoming divorced; (iv) more women are receiving more education.

Contrary arguments suggesting a stronger probability of a rising birth rate, for example above a completed family size of 2.3, are the anti-abortion campaign which loomed very large in the 1983 presidential election debate and the great acceleration of American economic growth. However, the anti-abortion campaign has not yet been translated into law or constitutional amendment. According to the fertility statistics available up to June 1984, the American economic boom has not so far been reflected in higher birth rates: the potential completed family size has not shown any sign of moving well above 1.9 since 1971. In any case the Census Bureau's line of reasoning is supported by other arguments: births to teenage mothers are falling; births to women over 30 are rising; first births to women over 30 are rising.

Therefore the population projections foresee faster demographic ageing than would a hypothetical choice of an even higher high fertility assumption at, say, a completed family size of 3. Is the present choice of assumptions well argued? The short answer is Yes. The points described above are well documented and there are no strong competing schools of thought.

Is it reasonable to expect that future fertility will follow the low variant? Historically, the fertility of the United States has remained slightly above that of many other North Atlantic countries even during the low periods of the Great Depression and of the 1970's. Perhaps this difference will continue but there are no clear contextual determinants of this difference. It may be due to the greater economic prosperity of the United States, or to the black American population with its fertility averaging just under 2,300 births per 1,000 women during the last 10 years, or to the greater proportion of regular church-goers in the United States. None of these explanations, nor the predictions that they imply, are supported by analytic studies. Therefore, it cannot be excluded that United States fertility falls to a potential completed family size of 1.6 children, following the example of several Soviet bloc countries in the 1960's or of several countries and regions of Central Western Europe in the 1970's.

Most importantly: Does the high fertility assumption chosen by the Census Bureau reverse the historic trend towards a rising proportion of elderly people? The answer is a weak Yes and a strong No. Table 6 shows that this high fertility trend would halt demographic ageing around the turn of the century much more substantially and for longer than the medium fertility assumption. The low fertility assumption does not halt the uptrend of demographic ageing at all. However, the reversal of the demographic ageing process by the high fertility trend is only temporary: all three fertility assumptions point to an upsurge in the proportion of elderly between 2005 and 2015. Obviously a swing back to the fertility of the postwar baby boom would prolong the reversal of the demographic ageing process indicated by the high fertility assumption in Table 6. Without such a swing back towards a plateau of total fertility rates of about 3,000 births per 1,000 women, the uptrend in the proportion of the elderly would be renewed, whichever of the three fertility assumptions is chosen, because the long wave of fertility between 1947 and 1970 will reach old age after 2012, definitely tipping the balance. Indeed the acceleration of demographic ageing towards 2005 or 2010 shown by Table 6 will be the very simple and very direct result of the gradual building up of fertility from 1939 to 1947, just before the high wave of the postwar baby boom.

In other words, it is not so much the low fertility assumptions and its projected trends which create the increase in the proportion of the elderly at the beginning of next century, but rather the postwar baby boom of 1947—1984 and, to a smaller extent, the slow climb of fertility out of the trough of the Great Depression which became quite steep between 1939 and 1947. The high fertility assumption's future trends are enough to reverse the demographic ageing process only until the force of the wartime "mini-peak" of fertility begins to be felt in 2008. Of course, when the full force of the postwar baby boom is brought to bear after 2012, the proportion of elderly would rise faster, overriding any see-sawing counteraction by the present high fertility assumption's future trends.

MORTALITY LEVELS AND TRENDS INTO THE FUTURE

Mortality is less problematic generally than either fertility or net immigration. Mortality analysis has a long history of increasing technical sophistication going back to the beginnings of demography three hundred years ago. While fertility analysis and the development of techniques for that task scarcely began before the twentieth century, mortality analysis was a serious pursuit in the second half of the nineteenth century owing to its importance for

life assurance. While the American demographic data-base was very poor in the last century, statistics for deaths by age, sex and medically certified cause have been reconstructed in an integrated series for ten Northern states and some other counties between 1900 and 1933. In this latter year, when all of the counties agreed to pool their certificates with the Federal government, the mortality statistics became available nationwide.

Mortality is a more settled subject also because it is intrinsically less complex than fertility or immigration. Death is a single event. At first glance, the arrival of an immigrant may seem to be a single event too, but emigration must be taken into account. The latter consists of re-migrating immigrants or of the native-born departing permanently. The definition of the permanence of emigration and the definition of immigration as distinct from tourism, student sojourns or family visits bedevil the measurement of net immigration. No such problems obscure the definition and measurement of death. Fertility is complicated in other directions. While mortality, like migration, can occur at any age and to either sex, fertility is more or less limited to ages 15-45. More important: women may have one or more children or none at all. Death is so simple by comparison: everybody dies but once; there is a diagnostic service legally required to explain each event.

It is for these sorts of reasons that past United States projections did not allow for more than one kind of level and trend of mortality. It is only the latest projections which present high, medium and low mortality variants. The presentation of three distinct levels and trends of mortality into the future in this case is a sign of increasing doubts about the possibility of simply extrapolating from past trends. There were no such doubts about following previous trend lines in the projections of 1977 and of 1974. What has happened recently to cause this degree of uncertainty?

Let us look at past developments. The great improvements in mortality demonstrated by the steady uptrend of life expectancy at birth in Figures 3A and 3B consisted almost entirely of gains before age 35, as infectious diseases were largely eliminated from the list of major killers by 1955.

This had been a long term trend since the nineteenth century. But, up to the base years of the latest projections, the gains made fell roughly into four periods. Between 1935 and 1955, the decline of central death rates accelerated an average of 1.7% per year for males and 2.6% for females. Previously, from 1900 to 1935, males had gained only 0.9% per year and females 1.0%. Quite unexpectedly and, indeed, still quite inexplicably, between the mid-1950's and

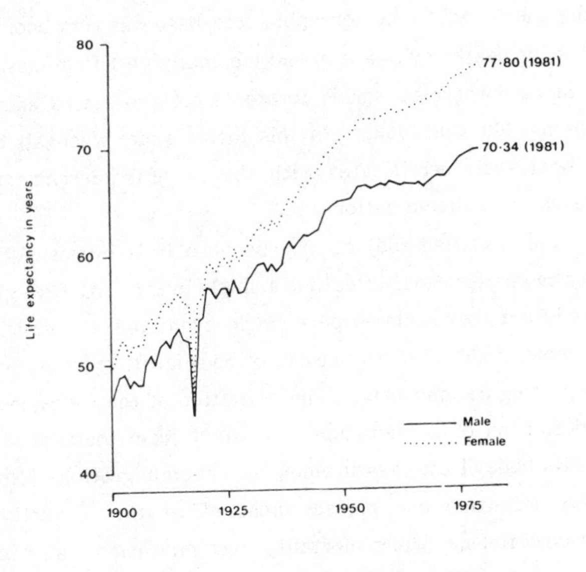

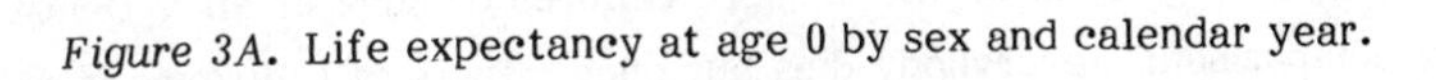
Figure 3A. Life expectancy at age 0 by sex and calendar year.

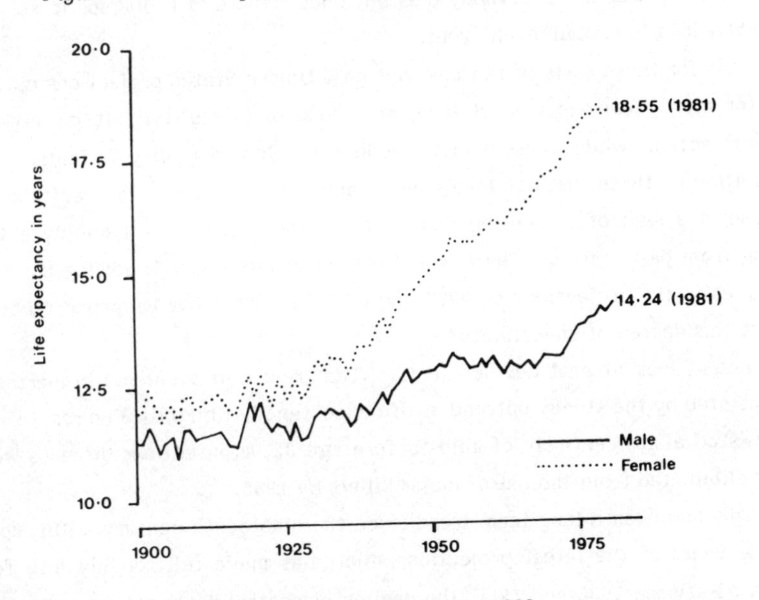

Figure 3B. Life expectancy at age 65 by sex and calendar year.
Source: US Social Security Administration, 1982.

the mid-1960's, males fell back by 0.1% of their initial central death rate each year. Female gains also decelerated but remained positive at 0.8% per year. From 1968 to 1981, by contrast, gains were faster: 1.8% for males and 2.2% for females, not far below the gains made in 1936-1954. Yet it is noteworthy that the female central death rate has not been rushing downwards so far ahead of the male rate in the recent period.

The years of 1900-1934 were a period of gradual expansion of public health improvements and slow progress in curative medicine. The second period, however, was typified by the introduction of highly effective, cheap, mass-produced curative medicines, namely the sulpha drugs and other antibiotics, as well as a wide range of vaccinations against infectious disease. In the middle of this period, moreover, there was the introduction of highly effective, cheap, mass-produced pesticides which reinforced the continuing extension of environmental health measures.

Altogether, as a result of the decline of infectious diseases, especially between birth and age 35, the probability of death at birth fell 90% between 1900 and 1980 for both males and females. At age 30 the decrease was 77% for males and 91% for females. At age 65 the probability of death fell only 30% for males but 61% for females between 1900 and 1980. At age 30 the sex difference was due, for the most part, to the persistence of high accident risks for males in contrast to the elimination of maternal mortality.

During the 1970's the mortality rates from cardiovascular diseases were reduced and, in turn, these brought down total male mortality at age 65 by 15% while female mortality, less affected by this cause of death, fell by only 9%. This is an example of male mortality converging on female mortality which is quite exceptional. On the whole, male:female mortality differences continue to diverge up to age 95.

Let us examine the age-specific improvements in central death rates which lay behind the above-discussed patterns for the whole population. Again focussing on the age brackets with the highest incidence of degenerative disease and the lowest incidence of infectious diseases at age 35-39, male mortality fell more than twice as fast in 1936-1954 as in 1900-1936, rose in 1954-1968 and then fell fairly rapidly in 1968-81. At ages 45-49, the sequence was practically the same except that faster gains were made in 1968-1981 than in 1936-1954. This is due to the more recent reduction in degenerative diseases. In the earlier period, advances in chemotherapy had a great effect in younger age brackets, not in middle-age. At ages 55-59, percentage improvements were not great in 1936-

54 because, from middle age onwards, infectious diseases never were the main element of mortality. However, mortality in late middle-age fell relatively sharply in 1968-1981.

At ages 65-69, gains in 1968-1981 were at more than twice the rate of 1936-54. However, at 75-79 and 85-89, the gains were approximately the same in both of these periods.

From these patterns, four hitherto unlearned lessons are evident: mortality has become more problematic since 1954; reversals of male life expectancy are possible; the degenerative diseases of middle and old-age are not irreducible; mortality after age 65 is flexible. These points contradict the accepted patterns and cast doubts on the apparently sound generalisations made by demographers 10 and 20 years ago. All of these points mean that the elderly population can now increase as a result of demographic processes after the age of 45. That is, since 1968, demographic ageing no longer depends so heavily on fertility but it took several years before this fundamental change was realised.

The female statistics do not show a reversal in the fall of mortality, but their down-trend did slow down from 1954 to 1981. Moreover, in 1936-1954 the rates of decline of mortality among women aged 50-74 were twice as fast as among males. That is, females did not have to wait for the advances in medicine and changing lifestyles that occurred in 1968-1981 in order to live longer beyond middle-age.

In sum, the average decline of the central death rate from 1900 to 1981 shows the greatest gains below middle-age. However, there have been great differences in the phasing of these gains with a swing towards some rapid reductions in male mortality in middle-age in 1968-1981. By contrast, female mortality in middle and old age fell before male mortality and there was no reversal of this downtrend among females in 1954-1968.

The comparison of the total percentage improvements in male mortality from heart disease, cancer and vascular lesions shows that these major killers of middle and old age generally continue to be reduced more slowly among males than females. However, Table 8 shows that there are important exceptions. These are important because they lend additional support to the real possibility of greater increases of life expectancy during middle and old age for males. They made relatively few gains in 1936-1954 and many losses in 1954-1968.

How do the Census Bureau's projections deal with the growing complexity and growing relevance of changes in mortality in late middle-age and old-age? The medium mortality assumption does not extrapolate the record gains of 1968-

Table 8. Negligible male:female differences in mortality decline, 1968-1978. (X = difference of 0.33% or less in annual reduction of central disease rate.)

Age	Five major killers by age bracket above 34				
	Heart Disease	Cancer	Vascular Diseases	Respiratory Diseases	Violence
35-39					
40-44			X		X
45-49	X		X		X
50-54				X	X
55-59	X	X		X	
60-64				X	X
65-69		X	X	X	X
70-74				X	
75-79					
80-84			X	X	
85-89			X		
90-94	X		X	X	

Source: US Social Security Administration, 1983.

1981 into the future. Instead it was decided, with very little and rather oblique explanation, that the recent spurt in life expectancy will not be sustained and that not even the low mortality assumption will continue the down-trend of 1968-1980. It is puzzling that, when for the first time the Census Bureau presents three alternative mortality assumptions instead of the customary one, the projections do not use the historic lows of mortality in 1936-1954 or 1968-1981. Moreover, from 2005, the three down-trends are slowed even further as Table 9 demonstrates.

Perhaps choosing a direct extension of the recent rapid down-trend would have been politically unacceptable, because it would empty the Federal Social Security pension fund. Of course, such a low mortality assumption, with all of its uncertainties, should not be taken as a serious threat to a fund which has worked for 50 years. A population projection should not be taken as a forecast. Rather it would have been helpful to see worked out, in detail, the implications of such a low mortality trend as did occur in 1936-1954 or 1968-1981. This would answer the crucial question: What if we continue for another fifteen

Table 9. Annual rate of increase of life expectancy as a percentage of the initial level

1968-1982	1982-2005 Mortality			2005-2080 Mortality		
	High	Medium	Low	High	Medium	Low
0.43	0.08	0.18	0.29	0.03	0.06	0.11

years the way we have been going for the last fifteen years? This would not imply that it will happen this way, but it would give a better sense of proportion for evaluating the medium and high mortality variants against the perspectives of the low mortality variant.

The reasons for choosing the three trend lines up to 2005 are not explained by either the Census Bureau or the Social Security Adminstration which provided the basic information and summary measures from which the Bureau worked. Nor is it clear why they chose to decelerate mortality decline once more after 2005 in each of the three assumptions.

It emerges from cross-checking the Social Security life tables against the puzzlingly cryptic Table 9 published by the Census Bureau that one may deduce the reason for the trends selected in the high, medium and low mortality variants. The high mortality assumption for 1982-2005 equals the percentage gains in life expectancy per year made in 1954-1968. The low mortality assumption equals the average percentage gains across the periods 1954-1968 and 1968-1982 combined. The medium assumption lies almost exactly in the middle.

Presumably the choices in Table 9 were argued from these recent levels and trends in this way. It would seem more reasonable to have used the average gains of 1954-1982 for the medium assumption, and the 1968-1982 experience for the low assumption.

What difference would these two completely new mortality assumptions have made to the projections of the proportion of elderly people?

To answer this, let us look first at the results of the high and low mortality assumptions when fertility and immigration are held at the medium level, as in Table 10. The high mortality assumption lowers the proportion of elderly substantially at the turn of the century, more than the medium mortality assumption and, of course, much more than the low mortality assumption. This low assumption does not produce any reversal in the uptrend of demographic ageing but, in this case, the increase of the elderly towards the end of the century is slow.

Table 10. Mortality and percent aged 65 and over: high, medium and low mortality assumptions

	Low mortality Medium fertility Medium immigration	Medium mortality Medium fertility Medium immigration	High mortality Medium fertility Medium immigration
1985	12.0	12.0	12.0
1990	12.8	12.7	12.6
1995	13.3	13.1	12.9
2000	13.4	13.0	12.7
2005	13.7	13.1	12.7
2010	14.5	13.8	13.2
2015	16.3	15.4	14.8

Source: US Bureau of the Census, 1984.

In the light of these contrasts between the mortality assumptions used in the 1982-based projections, it is evident that a new very low assumption drawn from the rapid decline of 1968-82 would tend to raise the proportion of elderly above the low assumption used in the 1982-based projection. Also, the pace of growth in the proportion of elderly would not almost come to a halt as it does with the present low assumption at the end of the century.

These regularities can be deduced because the Census Bureau's calculations use measures of mortality in each age and sex group of the population over the years to come which follow the patterns of mortality change described above in our review of developments since 1900, especially since 1968. Thus the lessons learnt recently about the new dynamics of mortality beyond age 35, especially among males in late middle age and old age, have been taken into account in choosing the *patterns* of future mortality, although the *levels* of mortality in 1954-1968 or 1968-1982 are not used directly to set the sights of future trends in the low assumption.

The report on the 1982 projections presents a great deal of background detail about the expansion of the proportion of elderly over age 74 and over age 84 which shows that their calculations did incorporate some of the lessons of 1968-1981.

Already in 1950, half of all deaths were after age 64; in 1980 two thirds. The medium projection points to a levelling off of the proportion of deaths in old age at 70-74% from 1985 to 2015. Instead, there is a shift upwards in the ages at which the elderly die. In 1980, almost a fifth of all deaths were at age 85 and over, more than double the proportion in 1950. In 2005, more than a quarter of all deaths will be after age 84, according to the medium projection.

The proportion of the elderly who are over age 84 had almost doubled between 1950 and 1980 from 4.8 to 9.1%. The medium projection foresees 15.1% by the year 2005.

These two developments mean that the proportion of the elderly will no longer increase only as a consequence of waves of children or waves of immigrants moving gradually up the population pyramid. This was the principal demographic ageing process until mortality rates between age 0 and age 35 fell close to zero in the 1950's, and the mortality rates of males as well as females became more dynamic in middle age and old age. Since mortality for middle-aged and old-aged males began to fall in 1968-1982, the entry of more and more people into old age depends to an increasing extent on mortality decline between age 35 and 65. In addition, more people reaching age 65 are proceeding past age 74 and age 84 solely due to falling mortality among the elderly.

The five year age- and sex-specific death rates used by the Census Bureau embody the new mortality dynamics into the projections, albeit without accepting the potentially very low level of mortality which would follow from the 1968-1982 down trend. In particular, the calculations take account of the slowing of female mortality decline at age 95 and over, with the consequence that male rates will tend to converge on female rates, bringing about a re-balancing of the male:female ratio at the very tip of the population pyramid in the long run. However, neither by 2005 nor by 2015 is there any prospect of the new mortality dynamics substantially changing the great preponderance of females between ages 65 and 95. While male mortality is projected to continue falling at these ages, female mortality is expected to fall faster on the whole, judging by experience in recent decades.

In order to concentrate on the processes connecting the three components of population growth with demographic ageing and their interactions, the present discussion avoids looking into the internal structure of the elderly population. However, the extremely skewed male:female ratio of the elderly and the mounting proportion of women who do not die within several years of passing age 65 cannot be forgotten. For these two outstanding features of the top of the population pyramid are due entirely to mortality, not to fertility or immigration. Now that it is the age brackets after 35 which bear the great brunt of mortality and now that its incidence has become increasingly dynamic between age 45 and age 85, mortality potentially has more direct and rapidly acting effects on demographic ageing than does either fertility or immigration.

Since the 1930's, both fertility and mortality have gone through great

changes of pace, shifted their age-specificity and gone up as well as down. Immigration may not have altered its age-specificity much, but it has moved in waves and tidal surges. Yet these swings in the three components - almost cycles - are not well-represented by the actuarial spelling-out of the more or less straight lines used in the American projections and indeed in most projections in North Atlantic countries. These swings will probably have great effects on the demographic ageing process especially after 2015. Therefore, there is still time to devise a more realistic projection method.

REFERENCES

U.S. Bureau of the Census (1984). Current Population Reports, Series P-25, No. 952. Projections of the Population of the United States by Age, Sex, and Race: 1983-2080. (by G. Spencer.) Washington DC: USGPO.

Fertility

Bloom, D.E. (1982). What's happening to the age at first birth in the United States? A study of recent cohorts. Demography, **19**, 351-370.

Butz, W.P. et al (1982). Demographic Challenges in America's Future. Santa Monica: Rand Corporation.

Devaney, D. (1983). An analysis of variations in U.S. fertility and female labor force participation trends. Demography, **20**, 147-161.

Easterlin, R. (1980). Birth and Fortune. New York: Basic Books.

Hendershot, G.E. & Placek, G.E. (1981). Predicting Fertility. Lexington: Lexington Books.

Jones, E.F. (1981). The impact of women's employment on marital fertility in the U.S., 1970-75. Population Studies, **35**, 161-173.

Masnick, G. & Bane, M.J. (1980). The Nation's Families: 1960-1990. Cambridge, Mass.: Joint Center for Urban Studies of MIT and Harvard University.

O'Connell, M. & Rogers, C.C. (1983). Assessing birth expectations data from the current population survey: 1971-1981. Demography, **20**, 369-384.

Preston, S.H. & McDonald, J. (1979). The incidence of divorce within cohorts of American marriages contracted since the Civil War. Demography, **16**, 1-25.

Smith, D.P. (1981). A reconsideration of Easterlin cycles. Population Studies, **35**, 247-264.

Immigration

Bean, F.D., King, A.G. & Passel, J.S. (1983). The number of illegal migrants of Mexican origin in the United States: sex ratio-based estimates for 1980. Demography, **20**, 99-109.

Davis, C., Haub, C. & Willette, J. (1983). U.S. Hispanics: Changing the Face of America. Population Bulletin 38. Washington DC: Population Reference Bureau.

Keely, C.B. (1977). Counting the uncountable: estimates of undocumented aliens in the United States. Population and Development Review, **3**, 473-481.

Siegel, J.S., Passel, J.S. & Robinson, J.G. (1980). Preliminary review of existing studies of the number of illegal residents in the United States. U.S. Immigration Policy and the National Interest: the Staff Report of the Select Commission on Immigration and Refugee Policy, Appendix E: Papers on illegal immigration to the U.S.

Warren, R. & Peck, J.M. (1980). Foreign-born emigration from the United States: 1960 to 1970. Demography, **17**, 71-84.

Mortality

Bourgeois-Pichat, J. (1978). Future outlook for mortality decline in the world. Population Bulletin of the United Nations, **11**, 12-41.

Crimmins, E.M. (1983). Implications of recent mortality trends for the size and composition of the population over 65. Review of Public Data Use, **11**, 37-48.

Fingerhut, L.A. (1982). Changes in Mortality among the Elderly: United States, 1940-1978. Vital and Health Statistics, Series 3. Washington DC: National Center for Health Statistics.

Keyfitz, N. (1978). Improving life expectancy: an uphill road ahead. American Journal of Public Health, **68**, 954-956.

U.S. Social Security Administration (1982). Life Tables for the United States: 1900-2050, Actuarial Study No. 87 (by J.F. Faber). Washington DC: S.S.A.

U.S. Social Security Administration (1983). Life Tables for the United States: 1900-2050, Actuarial Study No. 89 (by J.F. Faber & A.H. Wade). Washington DC: S.S.A.

Verbrugge, L.M. (1980). Recent trends in sex mortality differentials in the United States. Women and Health, **5**, 17-37.

Waldron, I. (1980). Sex difference in longevity, In: S.G. Haynes & F. Manning (eds.), Second Conference on the Epidemiology of Aging. Washington DC: National Institute of Health.

Wilkin, J.C. (1981). Recent trends in the mortality of the aged. Transactions of the Society of Actuaries, 33.

AGE STRUCTURE OF SOVIET POPULATION IN THE CAUCASUS: FACTS AND MYTHS

Zh. A. MEDVEDEV

Genetics Division, National Institute for Medical Research
Mill Hill, London, U.K.

INTRODUCTION

The significant increase in the proportion of elderly groups in most countries of the world which is closely related to the decline in the birth rate and the relative increase in lifespan, has revived interest in the geographical areas of high longevity and in societies and ethnic groups where ageing does not seem to be a social burden and where the old and the oldest seem to be useful for the life of the community. Among these areas, the most famous is the Caucasus - a highly populated mountain region situated between the Black and Caspian Seas. More than 40 national and ethnic groups which live there, with a total population well over 20 million, are officially classified as "longevous" or "long-living" people (Chebotarev, 1973; Chebotarev & Sachuk, 1980; Benet, 1976; Pitskhelauri, 1982; Brook, 1982). This definition is usually based on the claims of the unusually high proportion of centenarians among North Caucasian and Transcaucasian inhabitants (from 30 to 150 per 100,000, which is 10-30 times higher than in the rest of the Soviet Union or in Eastern or Western Europe).

The possibility of some natural environments in the world where for some reason or other people could live much longer and remain more vigorous in old age than in most industrial societies attracts attention not only from those who study the problems of ageing. The information about centenarians and "super-centenarians" is permanently exploited by all kinds of media and often misused for political and even for commercial reasons. Scientific reports about longevity are usually mixed with pseudoscientific "studies" and sensational discoveries of individuals living until 130, 140, 150 or 170 years old, women who gave birth to children when they are already 80 or 90 years old, or folk dance groups where all the dancers are centenarians.

In 1974 I made an attempt at a critical assessment of the Caucasus and Altay longevity and super-longevity phenomena (Medvedev, 1974). The last decade has shown that the whole problem remains controversial. Neither the

new Census in 1979, nor the two special joint Soviet-American symposia on "Biological, Anthropological and Social Studies of Longevity in the Caucasus" (1980 and 1982) were able to prove the scientific validity of the special status of the Caucasus as a geographical place exceptionally favourable for the prolongation of the human lifespan. Two approaches to the problem, enthusiastic and sceptical, still remain. However, the fact that tables of the age distribution of the Soviet population in different regions, normally published after each Census (in 1897, 1926, 1939, 1959, 1970) were not released after the Census of 1979, and official acknowledgment of the slight increase of the mortality rates in most parts of the USSR in 1976-1981 and reflected in all age groups of the population, make it possible now to consider the whole problem from a different perspective. In 1974, I placed most emphasis on the social and cultural factors which favoured the deliberate distortion of the real age structure of some ethnic groups, especially in rural areas of the Caucasus, Altay, Yakutia and some mountain places in Soviet Central Asia. Here I concentrate on the validity of the background statistical data and some results which were received after the attempts at verification of the official Census figures.

How valid is the Census information on centenarians?

The first comprehensive Census of population in Russia which, among many different indices, also registered the age distribution of population in all regions and areas was carried out in 1897. All information at that time had been based on verbal statements. The dominant rural population of Russia was not entitled to have internal passports or other valid identification documents. Only the urban population (about 10 per cent of the total in 1897) was entitled to have internal passports as some form of police registration documents. The possession of internal passports was a privilege; those without passports (peasants, Jews and some other national minorities) were not allowed to live in Moscow, St. Petersburg and many other towns and cities of Central Russia. The majority of the population of the Russian Empire was illiterate and this made the distortion of many patterns of the Census of 1897 inevitable. The Census of 1897 registered in Russia 15,677 centenarians (about 4,000 of them 100 years old). Urlanis (1978) calculated that in 1797 (100 years before the Census) the average birth rate in Russia was at the level of 990,000. This means that for every 100,000 born in 1796-1797 at least 400 were able to reach 100 years - the figure is clearly unrealistic. Taking into consideration the fact that infant mortality in 18th Century Russia was extremely high and life expectancy at birth

was well below 30 years old, the number of centenarians in Russia in 1897 seems to be seriously exaggerated. Critical analysis of this Census and the main causes of the inaccuracies in the age distribution demography were looked at by Novoselsky (1916). This is the main reason why in most Soviet publications on age-related demography, the materials of the Census of 1897 are not usually mentioned. The most recent publications in demographic gerontology compare the results of the Censuses of 1926, 1959, 1970 and 1979, which also show comparatively high proportion of centenarians (Table 1). These results are usually taken for granted and most Soviet experts have accepted without reservation the number of centenarians recorded from verbal statements.

Table 1. Reported number of Centenarians in the U.S.S.R.

	Census of 1926	Census of 1959	Census of 1970	Census of 1979
Total population	147,027,000[1]	208,826,650	241,720,134	262,436,227
Rural population	120,700,000	108,848,955	105,728,620	98,850,287
Age over 100				
Total	29,000	21,708	19,304	Not available
Men	12,000	5,432	4,252	
Women	17,000	16,278	15,052	
Rural		17,272	13,936	
Urban		4,436	5,368	

[1] within borders before 1939.

Sources: Itogi Vsesojusnoi perepisi naselenija (Census of 1970), Vol. 2, Moscow, Statistica, 1972, pp. 12-16; Naselenie S.S.S.P. (Population of the U.S.S.R.), Politizadat, Moscow, 1980, pp. 1-2.

The first serious critic of the validity of the number of centenarians shown in the Census of 1926 was Novoselsky (Novoselsky & Paevsky, 1933). He found that in the Census of 1897 oral statements about age showed a clear tendency of preference for round figures. In actual age groups there was always a maximal number for round ages (50, 60, 70, 80). The elevated figures were usual for figures divisible by 5 (45, 55, 65, 75, etc.) as well. Even numbers were usually slightly higher than odd numbers. The aberration was typical of rural, rather than urban populations, and occurred more frequently among illiterate groups and national minorities. The Census of 1926 showed the same aberration. Persons who were 58, 59 or 61 frequently reported their age as a round 60. This phenomenon, well known to demographers as "The Index of Accumulation of Ages

on Round Figures" was observed in other Censuses also and its prevalence was found to be higher in areas of greater illiteracy. This "Accumulation Index" can be calculated (normal statistical practice, forgotten for later Censuses) easily when the Census gives the age distribution for each annual group. A simple expression:

$$100 \times \frac{5(25+30+35+40) \ldots\ldots\ldots 70+75+80 \ldots\ldots)}{(23+24+25+26+27 \ldots\ldots\ldots 70+71+72 \ldots\ldots\ldots)}$$

where 23, 25, 30, 71, etc., are substituted by the number of persons registered for these ages is normally used to find the "Index of Accumulation in Round Figures". Novoselsky found that this index was 183 for the Census in Russia in 1897, and 159 for the Census in 1926 (Frenkel, 1949). For comparison, it is possible to indicate that the same index was 100 for the Census in Belgium in 1910, 100 for the Census in Sweden, 102 for Norway and Germany, which had Censuses in 1910 as well. For Censuses in 1911 in France and England, the factor of accumulation on round age figures was 106 and 107, respectively; 120 for the U.S.A. (the 20% age distortion was registered mostly among blacks and immigrants born outside the U.S.A.); 134 for Hungary and 137 for Spain (in 1910). The index was highest (248) for the Census in Bulgaria in 1905.

The aberration usually increases for older groups. For example, in the Census of 1926 in the U.S.S.R. the number of women 60 years old was registered as 1,147,000, while figures for 59 and 61 were 210,000 and 179,000 respectively (Frenkel, 1949). In more senior ages, the "Index of Accumulation" can reach 300, as can be seen in Table 2. Making some necessary corrections, Novoselsky and Paevsky (1933) calculated that the actual number of centenarians in 1926 was not 29,000, but well below 7,000. This means about 4 centenarians per 100,000 of the population, and an additional investigation could probably further reduce this "centenarian index". The Census of 1959 showed, as one would expect, a lower figure of centenarians, despite the significant growth of the total population (partly related to the post-war increase in the territory of the Soviet Union: addition of the Baltic states, Western Ukraine and parts of Poland, Bessarabia and some territories in the Far East). The Census of 1970 further reduced the "official" number of centenarians. At the same time, the *index of longevity* (the ratio of people 90 years and older to the number of 60-year-olds) accepted by Soviet gerontologists as a valid indicator of the "real" longevity in one or other region was also halved between 1926 and 1970, even for the Caucasus (Kozlov & Komarova, 1982).

Table 2. Distribution of elderly (over 78 years old) in the population of the U.S.S.R. according to the Census of 1926. The "accumulation" of population on round age figures: 80, 85, 90, 95 years (numbers in thousands)

Age	Male	Female
79	23	27
80	140	229
81	16	18
82	24	26
83	17	18
84	13	14
85	44	60
86	14	13
87	9	11
89	5	7
90	32	50
91	2	3
92	3	4
93	3	3
94	2	2
95	8	10
96	3	3
97	2	2
98	2	2
99	1	2
100 and over	12	17

Source: Frenkel (1949)

The decrease in the proportion of centenarians is usually explained by the improvement of statistical methods. Life expectancy increased during the same period from 44 to 70 years. It is difficult to calculate the "Index of Accumulation" for older groups in recent Censuses, because the published records do not give the figures of age distribution for each year, only for 5 or 10 year periods. It was reported, however, that the "Accumulation Index" was 102.5 for the Census of 1970 (Bolshakov, Brook et al, 1982) and that age pattern aberrations were mostly evident for older groups. The "Accumulation Index" was more obvious for the Census of 1959. The actual figure of centenarians registered in the Census of 1959 was later reduced by 22.5 per cent in some publications.

The number of centenarians revealed by the Census of 1979 was not published. There were only selective reports for Armenia (Kanzelson, 1983) and Abkhazia (Bolshakov et al, 1982). The latter verified the ages of centenarians in Abkhazia and reported that the official Census figure of 548 centenarians in

Abkhazia should be reduced to 241.

All Censuses (1926, 1959, 1970 and 1979) were based on verbal statements only and the subsequent verifications were made either on the basis of some statistical recalculations, or by a more direct expedition method (Bolshakov et al, 1982). Teams of experts tried to re-assess the official figures on the basis of personal interviews with already registered centenarians. Despite the ability of Bolshakov's team to reduce the number of centenarians in the most celebrated Caucasus region of Abkhazia from 548 to 241, the actual figure is probably much lower. The method of verification was still very arbitrary and not really scientific. The authors acknowledge the formidable difficulties of verification because of the absence of any records and because those who are 90 years old and older often "simply do not know how old they actually are". "Many of them (those older than 90) never went to school, could not read or write and could count only with difficulty. Such people constituted a majority in the areas of Abkhazia, Georgia proper and Azerbaijan which we have been studying. As for women, nearly all of them were like that. As a rule, birth certificates were non-existent in those areas. Among the Christians, registration of baptisms in church books was often burdened with mistakes. The names of the baptized babies entered into the books by the priests were not always used, and at home they were called by other names; besides that, Abkhazian women, after getting married, changed not only their surnames but also the first name. Many church books disappeared altogether.

As for the Muslims, they never practised registration of the newborn at all; only a few families marked the birth of a child by writing a line to this effect on the last page of the Koran. In everyday life people did not use calendar ages. They resorted to a system of age classes with rather vague boundaries. What is more, such classes were different for men and women (Bolshakov et al, 1982).

Bolshakov and his colleagues acknowledge that in their own study the verification of Abkhazian centenarians was very approximate. "It was based on an anamnestic method of recall; a conversation held with each one to clarify the main stages of life history, in particular in their family lives (the date of marriage, birth of children, etc.), to try and connect these events with the dated events of national or local importance. While working in Abkhazian villages the researchers used such milestones as the makhadzhir exodus (the migration of Abkhazians to Turkey in 1877), the "big snowfall" of 1911, the establishment of Soviet government, etc. The question about the big snowfall addressed to the

long-lived persons proved to be very effective. They were asked whether they remembered a time when many houses were buried in snow nearly up to their roofs. The additional question was whether the long-living subject helped senior members of the family to clear away the snow. As often as not such answers as, 'I remember the big snowfall very well, but I didn't help the adults, I was still a child,' or, 'Of course, I helped the adults, I was quite grown up then' immediately gave us an idea of the real age of those questioned (Bolshakov et al, 1982).

These simple questions made it possible for the authors to reduce the official Census figure of centenarians in Abkhazia from 548 to 241. But most other Soviet and foreign authors never tried any method of age verification - they just trusted the Census figures or repeated oral statements of "long-living" persons themselves.

The main limitation of Western authors who were studying and writing about longevity in the Caucasus (Leaf, 1973, 1975; Benet, 1976; Georgakas, 1980) was their inability to study valid documentation, as was done by Mazess & Formann (1979) for the Ecuadorian longevity case. The documentation was either not available or incomprehensible for foreign authors, who could read neither Abkhazian, nor the Georgian and Azerbaijanian languages. Therefore, all foreign studies were dependent on the information officially supplied by their Soviet colleagues. The main limitation of Soviet studies of the subject was the inability of Soviet authors to reconsider or to challenge the official statistical figures. Official statistical information is government-derived and the State Research Institutes (gerontology or ethnography) are not able (or do not wish) to receive funds and permission for substantial re-evaluation of official statistics, even for a specific group of supercentenarians in the 110-119 age group or those who are older than 120. The official Censuses are carried out not by qualified demographers but by ordinary persons, literate in local languages. These are usually students from local high schools or colleges, youth groups or people from offices. They fill in special forms, answering questions about names, place of birth, age, marriage status, educational background, language preferences, number of children, etc., on the basis of verbal answers. This method (with an inevitable margin for errors) well satisfies the Central Statistical Bureau, which is concerned with figures on population density, linguistic distribution, trends in migration, reserves of the work force in different regions, etc. For general statistical analysis it is important to know the number of old people of pensionable age (over 60 or 65). The number of centenarians is of no practical value for statistics related to the economy and the government

has no incentive to make an accurate assessment. However, in official government publications of Census information there was no attempt to personalise those who are "over 120", or to verify their exceptionally advanced ages. Those statisticians who finally assess the results of one or other Census are well aware of the possibility of false or exaggerated claims of super-longevity. There are quite a few Census forms, especially from less developed areas, like Altay or Yakutia, or mountain villages of the Caucasus, in which the age is indicated as 150, 170, 180 and even over 200. These forms are treated separately, and although these claimants are put into centenarian groups, they are in the section of "persons older than 100, for whom the exact age was not estimated". For the Census of 1959 this group consisted of about 10 per cent of all centenarians (2,182), while in the Census of 1970 these claims were simply dismissed as doubtful and were not included in the age distribution tables.

How valid is the case of the Caucasus longevity and super-longevity phenomenon?

The largest part of the Caucasus belongs to the Soviet Union. However, some parts of the Caucasus belong to Turkey and Iran. The mountain system extends over about 1200 km (750 miles). It is also about 500-700 km wide and contains two major inhabited areas: North Caucasia and Transcaucasia, with the main mountain range between them. In the Soviet part of Transcaucasia, the population is slightly more than 13 million (Census of 1979) and about 10 million in the North Caucasia. There are more than 40 nationalities in the Caucasus, with individual languages. The main nationalities are Georgians (Christians, about 3.5 million), Armenians (Christians, 4.1 million), Azerbaijanians (Muslims, 5.4 million). Other nationalities are mostly Muslims, except Osetians (0.5 million). They include Dagestanians (1.3 million; however, among Dagestanians there are 12 distinct ethnic groups with their own languages, Avarzy, Lezgins, Kymyks, etc.), Chechens, Kabardins, Ingushi, Kurds, Turks, Abkhazians, Adzharians, Balkarians, Cherkesses, and others. The climatic conditions are extremely different in different parts, some dry, some very humid. Food and national traditions are also different (with some common features). The most publicised longevity phenomenon is in Georgia, especially in the Muslim Autonomous Republic, Abkhazia. But this is simply because these regions were open to foreign visitors and are most easily accessible for study. In statistical terms, Soviet Azerbaijan has the highest proportion of centenarians. The claims of extreme longevity are typical for many Caucasus regions. It

Table 3. The age distribution among the elderly population in some areas of the U.S.S.R (numbers per 100,000)

Age groups	Ukraine		Latvia/ Estonia		Georgia		Azerbaijan		Karabakh-mountain region		Altay-mountain region	
	1959	1970	1959	1970	1959	1970	1959	1970	1959	1970	1959	1970
70-79	3243	3993	4984	5326	3612	3343	2464	2166	3905	3290	2575	2250
80-89	770	1252	1576	1821	1162	1383	987	943	1860	1998	914	1038
90-99	85	112	133	175	317	262	284	234	542	548	206	200
100+	5.7	5.4	2.6	2.6	51.3	39.3	83.8	48.3	144.4	99.7	47.3	28.5

Source: Census of 1970, Vol.2, Table 5. Statistica, Moscow, 1972.

would be appropriate to look at specific cases and to consider them separately. I do not plan to consider the longevity patterns in the many ethnic groups living in the Caucasus, but to concentrate attention on those which have been better studied.

The general proportional reduction in the number of centenarians (per 100,000) between the 1890s and 1970s was not random. The exceptional position of the nations of the Caucasus became clearer in 1959 than it was in 1897 or in 1926. However, among those living in the Caucasus the reduction of the number of centenarians was especially sharp between 1959 and 1970, as shown in Table 3. The figures for Armenia are much lower than for the other ethnic groups in the Caucasus (33.6 and 24.3 per 100,000 in 1959 and 1970) and are not included in the table. The Georgians and Azerbaijanians still record numbers of centenarians in comparison with other Soviet Republics and the highest figures are registered for the Karabakh mountain region in Azerbaijan. Abkhazians, the Muslim group in predominantly Christian Georgia, also show some decrease in the number of centenarians (78 and 60 per 100,000 in 1959 and 1970 according to the Censuses). However, one can find from the figures in Table 3 (which can be supplemented by the figures from many other regions of the Soviet Union), that general life expectancy is lower among "long-living" ethnic groups than among Ukrainians, Latvians, Estonians, or among the Russian population in the central areas of the Soviet Union. The mortality rates in Azerbaijan and Georgia are higher than in other parts of the European Soviet Union until age 75-79. Between 79 and 89, the mortality rate is still higher in Azerbaijan than it is in Latvia and Estonia, or in the Ukraine. Only after the age of 90 does the slow-down in the mortality rate become evident in

Georgia and Azerbaijan, mostly in rural areas of these national Republics. Among more than 10 different national and ethnic groups of the Caucasus there are several specific areas, like the Karabakh region in Azerbaijan and Abkhazia in Georgia where the numbers of centenarians are usually higher than average for the Caucasus. Within these areas there are some smaller districts where the figures of centenarians are even higher. Within these districts there are some special villages where the number of centenarians, according to the Censuses, are especially high, 5, 6 or more for the village. There are some famous villages where the centenarians claim the remarkable age of 120, 130, 140 or older. These villages, rather than the statistical figures for the whole regions, mainly contribute towards the publicity of the "Caucasus longevity phenomenon".

Table 4. The age distribution among centenarians in several areas of the U.S.S.R.

Region	Age group	Census of 1959	Census of 1970
Ukraine	100-109	1980	2010
	110-119	97	68
	over 120	4	4
Byelorussia	100-109	887	918
	110-119	51	37
	over 120	2	0
Latvia	100-109	71	73
	110-119	1	1
	over 120	0	0
		Caucasus	
Armenia	100-109	429	546
	110-119	72	40
	over 120	31	21
Georgia	100-109	1638	1614
	110-119	233	168
	over 120	105	62
Azerbaijan	100-109	2224	1985
	110-119	456	372
	over 120	279	128

Source: Census of 1970, Vol. 2, Table 5, pp. 257-262. Statistica, Moscow, 1972.

The official Soviet statistics based on Censuses do not consider super-longevity (ages over 120 years old) as unrealistic. This is reflected in the tables of age distribution *among centenarians* for all regions and nations of the Soviet Union published in the Census Reports. Tables 4 and 5 give a small selection of such statistics in order to make a comparison between the Caucasus and some other national groups. We have seen earlier that the number of centenarians per 100,000 almost halved between 1959 and 1970 in the Caucasus. However, for the "modest" centenarians (100-109 years old) the reduction of their numbers was not significant. The reduction in the numbers of those who claimed their age was either between 110 and 119, or over 120 declined more sharply and during rather a short period, despite significant improvement of the general medical service and special attention to the oldest groups just after 1958 when the newly created All-Union Institute of Gerontology in Kiev launched a special programme of medical assistance and studies of the long-lived people. Living standards in the Soviet Union also improved during this particular decade. It is, therefore, apparent that the reduction of the numbers of those who claimed to be older than 110 was a purely statistical phenomenon - the registration of "super-longevity" became more difficult and many claims were dismissed by local statistical bureaux and by the Central Statistical Bureau in Moscow as exaggerated.

It is also unusual that the sex distribution among centenarians, which as one may expect reflects the normal trend of the female dominance until 110, alters its character for more senior ages. For those "over 120" the number of males and females is approximately equal (Table 5), or even favours men in some areas of Dagestan, as shown in Table 6 (Alikishiev, 1962).

The age distribution among centenarians for 1979 was not published at all and this in itself is an important indicator. Soviet officials do not like negative trends in statistical records and the age-distribution statistics are not an exception. There was an increase in the mortality rates after 1975 for almost all age groups and life expectancy in the Soviet Union dropped in 1981 and 1982 compared with the 1960s and early 1970s (Tables 7-10). At the same time, the introduction of internal passports for the rural population in 1976, some other social improvements for the rural population pensions for members of collective farms, elimination of all taxes for land allotments for those older than 90 (*dolgozhiteli*, or long-lived persons), made local officials much more careful in accepting unchecked verbal statements about a person's age. If and when the factual data for 1979 or later years

Table 5. The age and sex distribution of centenarians in several rural areas of the U.S.S.R.

Region	Age groups	Census of 1959		Census of 1970	
		Male	Female	Male	Female
Ukraine	100-109	254	1157	196	1225
	110-119	15	50	7	37
	over 120	1	1	0	2
Georgia	100-109	534	840	464	862
	110-119	89	113	54	90
	over 120	43	59	17	33
Azerbaijan	100-109	667	1204	412	1004
	110-119	161	203	104	178
	over 120	92	133	42	53

Source: Census of 1970, Vol. 2,Table 5, pp. 257-262. Statistica, Moscow, 1972.

Table 6. Ratio of males and females among centenarians in one of the regions of Dagestan (Alikshiev, 1962).

Age group	Male	Female
100-104	22	50
105-109	21	26
115-119	10	7
130-139	4	2

are published, the exceptional status of the Caucasus as a land of special "long-lived" people will certainly be greatly reduced. But the Census figures as such cannot be taken as scientific by gerontologists. Censuses at which only verbal statements are registered are normally conducted for economic, demographic and political purposes. They include dozens of questions and indices about education, migration, languages, social status and many others. The accuracy of collected information is never perfect and it is less and less reliable for groups of people of more advanced age. It is certainly far from accurate in the specific group of centenarians and cannot be accepted without extensive additional study, which at first should verify the actual age of one or another person who claims to be 100 or 110, or 140 years old. Soviet research centres, like the Institute of Gerontology in Kiev, or

Institute of Ethnography in Moscow, or smaller centres, like the Georgian Centre on Gerontology, never considered any doubts about the official Censuses and their ability to give reliable figures of centenarians. There are many studies from these institutes which tried to find correlations between the relative number of centenarians and the altitude of their villages, the diets of local population, their fertility, working habits, etc. Nobody paid attention to the obvious fact that the largest proportion of centenarians, as a rule, correlates with the highest level of illiteracy, with the communities where there were not any birth or marriage registrations, and shows most clearly for small national minorities which did not have written languages.

The myth of the super-longevity phenomenon

The discussion in the previous section makes it clear that the available statistical information about centenarians based on the Census Report is unreliable for all practical purposes. Very little effort is necessary to show this. And yet, very few scientists are really interested in challenging the myths about longevity and super-longevity. The general public likes and easily believes the stories about longevity and super-longevity. The "discoveries" of individuals who have reached 120, 130 or 150 years of age make good newspaper, magazine or television stories. Even prominent gerontologists are often very enthusiastic about life extension possibilities by rather simple environmental factors. The famous Ilia I. Mechnikov, one of the first Russian promotors of life-extension methods, sincerely believed that the normal human maximal lifespan is somewhere near 200. Professor A. A. Bogomolets, organiser of the pioneering International Conference on Ageing Research in Kiev in 1938 and the author of the book "Prolongation of Life" (1938) also was convinced that the human lifespan potential is well over 150. He personally certified some long-living people in Abkhazia as aged between 130 and 140. Professor A. V. Nagorny, whose very serious monograph on ageing and longevity, published in 1940, was my first reading in gerontology, was even more enthusiastic. He was convinced that 200 years of age is not a real maximal potential limit. Quite a few other famous scientists in the recent past and present (I. P. Pavlov, L. Pauling, Albert Szent-Gyorgyi) were or are life-extension optimists with simple answers to all problems of super-longevity. For all these enthusiasts of simplified gerontology, the Caucasus gives a lot of examples of active, strong centenarians with record ages. The desire to be the oldest man in the world (Shirali Muslimov died in 1973 at the alleged age of 168) and the oldest woman in the world (Khfaf

Table 7. Life-expectancy of the male and female population of Russia according to Censuses (in years)

	Total population	Male	Female
Census of 1896[1]	32	31	33
Census of 1926[2]	44	42	47
Census of 1939	47	44	50
Census of 1959	69	64	72
Census of 1970	70	65	74
1978/79[3]	69	63.5	73.5
1981/82[3]	68	63.0	73.0

1 In 50 regions of the European part of Russia.

2 In the European part of the USSR within borders before 1939 (without Baltic states and Western Ukraine).

3 The figures until 1970 are reproduced from the data of the Census of 1979 which had been released (Volodarsky, 1983). The similar figures for 1978/79 were not published. They, as well as figures for 1981/82, were estimated by myself on the basis of the dynamic of the mortality rates per 1000 which had been published in the same book. The figures of life-expectancy in the Soviet Union for 1980/81; 61.9 years for males and 72.9 for females, published recently as an estimate by a Paris based National Institute of Demographic Studies (see Moynahan, 1983), are lower than one can calculate from the changes in general mortality rates.

Lasuria, died in 1975 at the alleged age of 142) is very addictive. After their death, new super-old were quickly discovered who are, of course, again the oldest people in the world. For local inhabitants, local scientists and for local politicans as well, the claim to fame of being the land of super-longevity is now very important and they publicise this through all possible channels. Local and central photographic agencies provide many Soviet and foreign publications with dozens of pictures of healthy and happy centenarians. In most capitals of the Caucasian Republics (Tbilisi, Baku, Erevan, Makhach-Kala and others) there are sections on local long-lived people in the Museums of National History. There are postcards and photoalbums for sale on the same

Table 8. The birth and mortality rates in Russia and the Soviet Union in 1913-1982 (per 1000 of population)

Year	Born	Died
1913	45.5	29.1
1926	44.0	20.3
1940	31.2	18.0
1950	26.7	9.7
1960	24.9	7.1
1970	17.4	8.2
1971	17.8	8.2
1972	17.8	8.5
1973	17.6	8.7
1974	18.0	8.7
1975	18.1	9.3
1976	18.4	9.5
1977	18.1	9.6
1978	18.2	9.7
1979	18.2	10.1
1980	18.3	10.3
1981	18.5	10.2
1982	19.0	10.1

Source: Volodarsky, 1983.

Table 9. Age-specific death rates in Russia and the Soviet Union: 1896 to 1976 (per 1000 in each age group)

Age groups	1896/ 1897	1938/ 1939	1958/ 1959	1965/ 1966	1969/ 1970	1970/ 1971	1974/ 1975	1975/ 1976
All ages	32.4	17.4	7.4	7.3	8.2	8.2	9.0	9.4
0-4	133.0	75.8	11.9	6.9	6.9	6.7	8.2	8.7
5-9	12.9	5.5	1.1	0.8	0.7	0.7	0.7	0.7
10-14	5.4	2.6	0.8	0.6	0.6	0.5	0.5	0.5
15-19	5.8	3.4	1.3	1.0	1.0	1.0	1.0	1.0
20-24	7.6	4.4	1.8	1.6	1.6	1.6	1.7	1.7
25-29	8.2	4.7	2.2	2.0	2.2	2.2	2.1	2.1
30-34	8.7	5.4	2.6	2.6	2.8	2.8	3.0	3.0
35-39	10.3	6.8	3.1	3.2	3.7	3.8	3.7	3.8
40-44	11.8	8.1	4.0	3.9	4.7	4.7	5.2	5.3
45-49	15.7	10.2	5.4	5.1	6.0	6.0	6.7	6.9
50-54	18.5	13.8	7.9	7.9	8.7	8.7	9.0	9.3
55-59	29.5	17.1	11.2	11.1	11.7	11.8	13.0	13.4
60-64	34.5	24.5	17.1	17.2	18.0	17.9	18.3	18.9
65-69	61.6	35.1	25.2	25.2	27.5	26.9	27.4	28.0
70+	89.0	78.9	63.8	65.8	75.7	74.9	73.3	75.0

Sources: Nar.khoz. 1922-72, p. 43
Nar.khoz. 72, p. 49
Nar.khoz. 74, p. 47
Nar.khoz. 75, p. 43

1975/76: Davis & Feshbach (1980)

subject. Before Shirali Muslimov was discovered as the oldest man, the record belonged to Mukhamed Eivazov, 147 years old. A special postage stamp was issued in 1956 to commemorate this. (He died three years later, at the alleged age of 151, but in 1963 Shirali Muslimov was discovered and publicised as an even older man of 158 years.) There were numerous articles about him during the next decade in most Soviet newspapers and in the West. "Life" magazine published a long article about Muslimov (Young, 1966) with many photographs. On one of the pictures, Muslimov was with his third wife, Khatun, aged 90. Shirali told the U.S. reporter that he married her at the age of 110. He also said that his parents had lived to over 100 and that his brother had died at the age of 134. When an American journalist wanted some valid information about Muslimov's age, he was not offered any, except the Muslimov's grandson (a local doctor) who swore "that the old man was given a small copper plate at birth inscribed with his name and the date". This plate subsequently was lost, but it was lost after Muslimov was already a celebrity. Several Soviet and foreign films had been made about Muslimov.

The local and central press have used all occasions possible to publicise the longevity achievements. When the oldest woman, Khfaf Lazuria, celebrated her 140th birthday, all Soviet newspapers published articles about it. In the Moscow *Izvestia* (12th October, 1974) one could read the vivid story of

Table 10. Age- and sex-specific mortality rates in the U.S.S.R. in 1973/74 (number of deaths per 1000 for each age group)

Age groups	Total	Male	Female
All population	8.7	9.3	8.2
0-4	7.7	8.5	6.8
5-9	0.7	0.8	0.5
10-14	0.5	0.6	0.4
15-19	1.0	1.4	0.6
20-24	1.6	2.5	0.8
25-29	2.0	3.1	0.9
30-34	2.8	4.4	1.4
35-39	3.6	5.4	1.8
40-44	4.9	7.4	2.6
45-49	6.4	9.7	3.7
50-54	8.8	13.9	5.8
55-59	12.3	19.5	8.2
60-64	18.2	28.7	12.6
65-69	27.0	40.9	20.2

Source: Bedny, 1978, p. 28.

how Lazuria's old friends came from different places "to congratulate Lazuria, 120-year-old Foma Shlarba, 100-year-old Ediche Aichiba, and Niko Djonua. Astan Slarba, who is 111 years old, came galloping - his village is 30 kilometers from Kutol"......."Khfaf Lazuria continues to work in her kolkhoz .... in 1973 she collected 2000 kilogram of tea leaves, a quota which not every young farmer can fulfil."

The long-living could have special economic incentives. When a dance group of centenarians was organized in the Abkhazian village of Duripsh (the oldest was Rakhaimi Butba, 115 years old), the village became a tourist attraction. The local government built a special club there, a hotel for guests and visitors, and new houses for many inhabitants (*Pravda*, 22nd January, 1976). Khfaf Lazuria died about one year later than Muslimov, on February 14th, 1975. Her death was reported as a world news item. It was said to have been an accident, she got pneumonia after swimming in a local stream during winter. But she still had some strength to visit the local cemetary and show the place where she wanted to be buried.

After Muslimov's death (there were several versions of its cause) the oldest man was Shirim-baba-Gasanov. He lived nearby and was reported to be 156 years old in 1973. Like Muslimov he was a Talysh, which is a small ethnic group of several thousand people with their own language. Gasanov died two years later. This reduced the age of the oldest man to 143; this record belonged in 1977 to Medzhid Agaev (*Pravda*, 23rd July, 1977). He died on November 14th, 1978, soon after a special film had been made about his life. *Izvestia* reported recently (March 18th, 1984) that a special museum has now opened in Lernika, centre of the Kelbadzhar district where all three famous long-living men spent their lives.

In 1981, the longest living man and woman were still Azerbaijanians, Yadigyar-Kaslhi (135 years old in 1981) and Gulyandam-Gary, the oldest woman, 144 years old in 1981 (Ivchenko, 1981). However, Gulyandam-Gary probably died in 1981 or 1982. The oldest woman in 1982, according to *Pravda* (11th April, 1982) was Apruz, from the Azerbaijan village of Kirovsk. She was 141 in 1982. There are, of course, dozens of centenarians in the Caucasus who claim to be between 120 and 135, but it will be a long time to wait for one of them to reach the age of these legendary men of the 1970s.

ACKNOWLEDGMENT

The author thanks Ms. Helen M. Crowne and Ms. Daphne Field for assistance in editing and typing the original typescript of this paper.

REFERENCES

Alikishiev, R.Sh. (1962) Longevity in Dagestan (in Russian), In: V. V. Alpatov (editor), Problems of Gerontology, pp. 16-21. Moscow: Publishing House, Academy of Sciences of the USSR.

Bedny, M. (1978). Demography and biology of ageing (in Russian), In: D.J. Vallentai (editor), The Health of the Elderly, pp. 26-38. Moscow: Statistica Publishing House.

Benet, S. (1976) How to Live to be 100. The Life-style of the People of the Caucasus. New York: The Dial Press.

Bogomolets, A.A. (1938) The Prolongation of Life (in Russian). Kiev: Publishing House of the Ukrainian Academy of Sciences. English translation, 1947, New York: Duell Sloan & Pearce.

Bolshakov, V., Brook, S. & Kozlov, V. (1982). Specific features of the statistical studies of longevity. Paper presented at the Second US/USSR Symposium on Longevity: Research in the Caucasus, November 1982, New York. (Collection of papers of the Symposium is deposited at the National Institute of Aging (NIH); not yet published.)

Brook, S.J. (editor) (1982). The Phenomenon of Longevity (in Russian). Papers of the First US/USSR Symposium on Longevity (Tbilisi). Moscow: Nauka Publishing House.

Chebotarev, D.F. (1973). Longliving and its role for understanding of ageing (in Russian), In: D.F. Chebotarev (editor), Longevous People, pp. 5-9. Kiev: Gerontology and Geriatrics Year Book.

Chebotarev, D.F. & Sachuk, N.N. (1980). Ageing. Union of Soviet Socialist Republics. In: E. Palmore (editor), International Handbook on Ageing, pp. 401-417. London: Macmillan.

Davis, C. & Feshbach, M. (1980). Rising Infant Mortality in the USSR in the 1970s. U.S. Department of Commerce, Bureau of Census (Series P-95, No. 74). Washington DC: Government Printing Office.

Frenkel, Z.G. (1949). Prolongation of Life and Active Ageing (in Russian). Moscow: Academy of Medical Sciences of the USSR.

Georgakas, D. (1980). The Methuselah Factors. The Secrets of the World Longest-lived Peoples. New York: Simon & Schuster.

Itogi Perepisi Naselenija SSSR (Census of 1970) (in Russian), Vol. 2, 1972. Moscow: Statistica Publishing House.

Ivchenko, V. (1981) Ageing never comes here (in Russian). Vokrug Sveta (Around the World), Moscow, No.7, 46-52.

Kanzelson, Y. (1983). The secret of long life. Interview with Minister of Health of Armenian Soviet Republic. Novosty Press Report, released on 7th August, 1983. Erevan/Moscow: Novosty Press.

Kozlov, V.J. & Komarova, O.D. (1982). Geography of longevity in the USSR (in Russian), In: The Phenomenon of Longevity, edited by S. J. Brook, pp. 31-40. Moscow, Nauka Publishing House.

Kozlov, V.J. & Komarova, O.D. (1982). Geography of longevity in the USSR (in Russian), In: S. J. Brook (editor), The Phenomenon of Longevity, pp. 31-40. Moscow: Nauka Publishing House.

Leaf, A. (1973). Getting old. Scientific American, **229:** 45-52.

Leaf, A. (1975). Youth in Old Age. New York: McGraw-Hill.

Moynahan, B. (1983). Article in The Sunday Times (London), 25 September 1983.

Mazess, R.B. & Forman, S.H. (1979). Longevity and age exaggeration in Vilcabamba, Ecuador. J.Gerontology, **34:** 94-98.

Medvedev, Zh. A. (1974). Caucasus and Altay Longevity: a biological or social problem? The Gerontologist, **14:** 381-387.

Nagorny, A.V. (1940) Problem of Ageing and Longevity (in Russian). Kharkov: Kharkov State University Press.

Narodnoye Khoziaistvo SSSR (1972, 1973, 1974 etc.) (in Russian). Statistical Yearbook of National Economy of the USSR. Moscow: Statistica Publishing House.

Novoselsky, S.A. (1916). Mortality and Life Span in Russia (in Russian). Petrograd.

Novoselsky, S.A. & Paevsky, V.V. (1930). The Mortality and Life Span of Population in the USSR (in Russian). Moscow.

Novoselsky, S.A. & Paevsky, V.V. (1933). A question of recalculation of the age-groups population numbers (in Russian). Proceedings of the Demographic Institute of the Academy of Sciences of the USSR, vol. 1. Moscow.

Pitskhelauri, G.Z. (1982). The Longliving of Soviet Georgia. New York: Human Science Press.

Volodarsky, L.M. (editor) (1983). Population of the USSR (in Russian). Moscow: Politizdat.

Urlanis, B.Z. (1978). The Evolution of Lifespan (in Russian). Moscow: Statistica Publishing House.

Young, O. (1966). In the mountains of Azerbaijan life goes on and on. Life Magazine, 16 September 1966, 121-127.

THE HEALTH OF AN AGEING POPULATION

J. GRIMLEY EVANS

Division of Geriatric Medicine,
Nuffield Department of Clinical Medicine,
The Radcliffe Infirmary, Oxford, U.K.

The structure of an ageing population such as that in Britain has been determined largely by fluctuations in the birth rate and the downward trend in infant mortality; the impact of improvements in mortality in adult life has been small (Benjamin, this volume). It is, therefore, possible to make accurate minimal estimates of the numbers of the elderly in Britain over the next 20 years: given only moderate improvements in mortality, we can expect little or no change in the overall numbers of those aged 65 and over, but a 16% increase in those aged 75 and over and a 48% increase in those aged 85 and over (O.P.C.S., 1983). We shall use the British population for illustrative purposes, but human ageing is a universal process and population ageing is a phenomenon about which nations can learn through the experience of others. Perhaps nowhere is the issue more vivid than in the dilemmas facing the policy-makers of China who seem to have achieved the impossible in rigorous control of the birth rate (Keyfitz, 1984). The future well-being of the Chinese nation depends on the skill with which its population structure can be titrated against its resource production and consumption. If successful, this endeavour will truly represent a giant step for mankind in rational environmental control.

Age-specific incidence rates for some important age-associated diseases follow a power-law relationship to age. This is so for most adult cancers, although the curves for cervix uteri and breast cancers in the female show a point of inflection towards a less steep increase with age in later life (Doll, 1970). Incidence rates for cerebrovascular stroke also follow a power-law relation against age (Evans & Caird, 1982), as do mortality rates from arteriosclerotic heart disease, although in this case there is a change of slope in the male about the age of 50 (Evans, 1984). Burnett (1974) suggests that the power-law relation of a disease to age reflects "some combination of self-activating processes of intrinsic origin". In the case of cancers, however, epidemiological

evidence is convincing that at least 70% are extrinsically caused (Doll & Peto, 1981), while Peto et al (1975) demonstrated that a power-law relation to age could come about as the result of cumulative exposure to an environmental cause. It is therefore tempting to speculate that those diseases showing a power-law relation to age have important extrinsic determinants, and there is collateral epidemiological evidence to support this view for both stroke and coronary heart disease.

If we turn from the incidence of specific diseases to the prevalence of disabilities, or the functional capacity of the human organism as a whole, we more commonly find an exponential relationship with age. This is approximately true over the latter part of adult life for age-specific mortality rates, the prevalence of chronic disabling disease (O.P.C.S., 1973), institutionalisation rates revealed in Census data, general practitioner consultation rates (O.P.C.S., 1974) and length of hospital stay in the Hospital In-Patient Enquiry. Although these indices increase exponentially with age the steepness of increase varies considerably under the influence of factors determining the transition probabilities between particular functional decrements and categorisation by diagnosis or use of resources.

Table 1. Age factors (ten year proportional increase) in prevalence of disability in elderly people at home.

	Age factor
Difficulty with household tasks	2.9
Inability to do shopping	3.2
Inability to get out of doors	4.4
Difficulty with self care:-	
Cutting toe nails	2.0
Bathing	2.9

As a broad index of the slope of the exponential function against age the ratio increase over 10 years of life is intuitively more meaningful than a more mathematical approach. In Tables 1 to 3 we present some of these age ratios for a variety of measures of morbidity and use of Health and Social Services by the elderly population. With the ageing of the population, assuming that present patterns of morbidity and use of resources continue into the future, the impact of population ageing will be greatest for those

factors showing the largest age ratios. Clearly the assumption that patterns of morbidity and use of resources will or should remain the same, is where our scope for fashioning the future lies.

Table 2. Age factors (ten year proportional increase) in use of certain health and social services

	Age factor
Home help	4.0
Meals-on-wheels	5.3
District Nurse or Health Visitor	3.2
Chiropodist	2.3
General Practitioner:-	
All consultations	1.2
Home visits	2.3

There are undoubted irrationalities in the use of resources identified in Table 2. The failure of the meals-on-wheels service to meet nutritional or social needs now seems well established (Johnson et al, 1982) and it would be foolish for us to envisage the expansion of this service to cope with the change of population over the next 20 years on the present pattern of deployment. Other approaches to the nutritional and social needs of the elderly would be more effective. It is likely that some of the slopes in the tables are steeper than they should be because the purveyors of health and social services react to the subject's age rather than to an objective assessment of need. One suspects that a lady of 80 will be more likely to be given a home help than will a younger applicant even if their levels of disability are the same. Conversely, the failure of the use of general practitioner services to increase as steeply with age as do other indices (Table 2) might well reflect inadequacies in primary care for the elderly. But similar reasoning may apply to some of the use of hospital resources for acute illnesses (Table 3). The slope in the use of high technology medical care against age has perhaps been kept down in Britain by discrimination against the elderly in access to modern methods of treatment. Evidence is accumulating, however, that older people may have significant benefits in terms of improvement and quality of life from modern medical technology and that it is inhumane as well as socially unjust to deny them such benefit. Treatment of chronic renal failure (Taube et al, 1983), operative approaches to coronary heart disease, and other high technology aspects of modern medicine may be expected to join orthopaedic

Table 3. Age factors (ten year proportional increase) in hospital admissions and use of hospital beds by various categories of disease

	Hospital admissions	Use of hospital beds
	Age factor	Age factor
All forms of vascular disease	1.9	3.5
Cerebrovascular accidents	2.3	3.9
Musculoskeletal disorders	1.2	2.9
Cancers	1.1	1.5
Respiratory disease	2.0	2.9
Proximal femoral fracture	3.5	5.0

Table 4. Age factors (ten year proportional increase) in prevalence of defined diseases

	Age factor
Arthritis of hip	1.5
Stroke	1.6
Dementia (moderate or severe)	4.4
Eyesight difficulties*	1.7
Hearing difficulties*	1.4

*People living at home only.

joint replacement as procedures with a definite place in the management of disease and disability in old age.

Most significant and alarming in its implications is the steep increase with age shown in Table 4 in the prevalence of dementia. This is a pervasive disability which in addition to the needs which it generates directly for medical and social care also modulates the need for care in a variety of other conditions. Most elderly patients in English long-stay geriatric hospitals, for example, are there because they are demented as well as being physically disabled. Dementia differs from most physical disabilities in requiring continuous care and this is the type of care which cannot realistically be provided in domiciliary settings (Isaacs

& Neville, 1976). Planning to cope with the expected increase in elderly people with this disability has therefore to be of a very different type from planning for the physically disabled. Unfortunately, there is no evidence at present in Britain that the policy makers and administrators of the Health and Social Services have assimilated this fact.

Conclusions drawn from Tables 1 to 4 will be affected by changes in the fatality of diseases and disabilities and the effects of treatment. Fries (1980) has drawn attention to the 'rectangularisation' of the population survival curve over the last century and has prophesied that in advanced societies environmental effects upon mortality will shortly be reduced to a minimum and that intrinsic ageing will lead to rapid death at an age with a limited variance around a species-defined mean. This view, suffused as it is with New World optimism, is an unfortunate hypothesis in that it may encourage health service administrators in the belief that the problems of an ageing population will solve themselves and others in the belief that good quality medical care is not necessary for older people because they are suffering from intrinsic and hence uncontrollable senility. It also depends on a number of questionable assumptions. The central problem is that the Fries model, however well it might apply to mortality, takes inadequate notice of morbidity: also, we do not know what the mechanisms of intrinsic death are likely to be. The assumption that as the survival curve is rectangularised by mortality the period of morbidity and disability preceding death will also grow shorter is questionable. Indeed there is evidence that the time between the diagnosis of senile dementia and death has been increasing in recent years (Blessed & Wilson, 1982) and this factor adds greater weight to the already very alarming effect of the ageing population on the prevalence of dementia implicit in the figures in Table 4. So far the only evidence that senile dementia of the Alzheimer's type has extrinsic determinants comes from a hint of a decline in incidence of senile dementia in a Swedish population (Hagnell et al, 1981). This finding may possibly be related to that of another Swedish study which suggests that successive cohorts of Swedes are reaching old age with greater reserves in mental faculties (Berg, 1980). If the overt manifestation of dementia are, as some have suggested, a threshold effect determined by initial abilities as well as by the pathological destruction of neural tissue, the age of manifestation of dementia may conceivably be becoming later in Sweden, even though the incidence of the pathological process has not changed. There appears to be no evidence from any other country of any

Table 5. Differences between young and old

True ageing:		
	(1)	Intrinsic
	(2)	Extrinsic
Cohort effects		
Aggravated ageing		
Secondary ageing		

improvement in the appalling toll that dementia takes on the elderly population and there is certainly no sign in Britain to discourage us from the view that dementia will dominate and may seriously impair our system of health and social services unless we take appropriate action for the future.

In looking for ways to take control of the future, we should consider not just the deployment and effectiveness of services but also the determinants of age-associated disease and disabilities. This can be approached in the context of a general epidemiological model of ageing, summarised in Table 5, which sets out the origins of differences between young and old people as observed in cross-sectional studies of population samples. Ageing is most usefully characterised as loss of adaptability of an individual organism with time. This loss is the result of intrinsic (inherited, mainly genetic) factors and extrinsic (environmental) factors and their interactions. But true ageing, in the sense of changes that have occurred to an individual as he grows older, is only one of the sources of difference between young and old people. Cohort differences are particularly prominent in rapidly changing societies such as our own. The mental skills of the elderly are different from those of younger people brought up in a different educational system and in a social group with different values, but these cultural differences may be misinterpreted as ageing changes (Schaie & Strother, 1968). Although most dramatically characterised in the study of psychological function, cohort effects have important effects on physical status, notably in the consequences of cigarette smoking (Svanborg, 1983).

If ageing is characterised by loss of adaptability it can only be assessed by presenting people at different ages with similar challenges and measuring their response. In some important ways society is so organised that older people, and particularly impaired older people, may be presented with more severe challenges than face the young. The consequences of such 'loading of the dice'

against the elderly are included in Table 5 as 'aggravated ageing'. In Britain we aggravate true age-associated impairments in body temperature control in the elderly (Collins et al, 1977) by a housing policy which places the elderly preferentially in poorer and colder housing (Fox et al, 1973; Wroe, 1973). In Newcastle we demonstrated that mentally impaired old people with proximal femoral fractures were more likely than mentally fit elderly people with the same injury to be sent to a hospital which provided poorer standards of care (Evans et al, 1979,1980).

Table 5 adds to a purely epidemiological picture of human ageing by including the concept of secondary ageing, processes of adaptation to primary ageing effects. One of the basic biological enigmas of ageing, to which Kirkwood (this volume) offers an answer, is why living organisms characterised by capacities for self-repair cannot or do not make good the ravages of age. Some adaptive and compensatory processes may however be activated. At the level of the individual, psychological adaptations to impairments are obvious enough. Helderman et al (1978) have also suggested that the age-associated increase in sensitivity of osmoreceptors is an adaptation to renal impairment. Such secondary ageing adaptations are of some general interest. Homeostatic mechanisms typically comprise nested servoloops in which loops with short time-bases work in fields defined by those with longer time-bases. Failure to repair ageing damage might be due to the absence of repair mechanisms or the absence of a sensor mechanism able to detect deterioration in function occurring very gradually over a long period of time. In the latter case repair mechanisms might possibly be activated artificially. The existence of secondary ageing adaptations suggest that, for some systems at least, absence of repair lies with the effector rather than the sensor mechanisms.

Somewhat more contentiously, the human menopause may be an evolved adaptation to age-associated impairment in the efficiency of female reproduction in the unique context of a species with a family social structure and, through speech, a cumulative culture. The menopause takes place at that age when in the past it became biologically more effective for the human female to give up increasingly unsuccessful attempts to produce viable offspring of her own, even though each of these contained 50% of her genes, and to contribute instead to the survival of her grandchildren each containing 25% of her genes. The male's biological investment in unsuccessful pregnancies is so much less that there would be little selection pressure to favour total infertility

for him at later ages. This interpretation requires a number of assumptions including the uniqueness of the human menopause. This is a matter of some debate, but one is not impressed with the senile changes seen at advanced ages in some (but not all) monkey ovaries (Graham et al, 1979) as an analogue to the menopause. It also depends on survival beyond the age of the menopause having been a frequent occurrence at an early stage in human evolution. The present uncertainties surrounding archaeological demography (Molleson, this volume) prevent any firm view on this possibility.

In a rational response to the challenge of an ageing population we clearly have an urgent need to identify and ameliorate extrinsic factors in ageing. Essentially there are two methods, the interventive and the observational. In the interventive approach the environment of individuals is changed and their response assessed. An example is the experiment by Aniansson and Gustafsson (1981) showing that retraining programmes for elderly men improved physical fitness and changed the histology of muscles towards a younger pattern. This approach will only detect the labile component of extrinsic ageing and it is to be expected that much extrinsic ageing is irreversible; the accelerated ageing of the lungs of cigarette smokers is a case in point. The observational approach measures and compares age-associated changes in people living under different environmental conditions. This traditional epidemiological approach, correlating environmental variables or other risk factors with measures of morbidity, encounters some particular problems when applied to ageing. If applied only to the elderly it may seriously underestimate the contribution of extrinsic processes to ageing. There are several reasons for this but probably the most fundamental lies in the use of current rather than past or cumulative measures of exposure to identified extrinsic factors. Our own studies of stroke, for example, provided evidence that although obesity in old age is not a predictor of stroke a history of having been obese in early adult life apparently is (Evans, 1983). There was also a hint in our data that hypertension early in life may be a better predictor of stroke in old age than hypertension late in life. Possibly the same mechanism may underlie the finding of the Birmingham Stroke Study that a history of diagnosed diabetes increased the risk of stroke but demonstrable glucose intolerance did not (Peacock et al, 1972).

Recognition of changes in incidence of age-associated diseases over time is a special case of the observational method for identifying extrinsic effects of ageing, and important for predicting future service needs. There are some important age-associated diseases which are currently showing

extrinsically determined changes in incidence. On the credit side it now seems well established that the incidence of cerebrovascular stroke is falling (Garraway et al, 1979). Mortality rates have been falling in many Western countries for many years, for reasons that are not entirely clear, although the rate of decline in the United States probably accelerated since programmes for the effective control of hypertension in the community have been widely disseminated (Ostfeld, 1980). The incidence of coronary heart disease in the United States has also been falling (Havlik & Feinleib, 1979) and mortality may now be falling in this country too (Marmot, 1984). In some countries such as Sweden mortality is still rising (Welin et al, 1983).

Less encouraging are trends in proximal femoral fracture in the elderly. In Britain this is a highly important disease accounting in some centres for as much as 50% of acute orthopaedic beds. It has been suspected for some years that the incidence of this disease is rising but the studies claiming this have either involved routinely generated hospital data (Lewis, 1981), which can be misleading due to a number of artefacts or have not been clearly related to a defined population and properly adjusted for age. A series of epidemiologically based studies in Newcastle have suggested that there probably has been an increase in incidence of this disease in older women but that the increase is not as great as is implied by the less epidemiologically rigorous studies.

The reasons for the increase are not clear. Svanborg (1983) has suggested that it is due to cigarette smoking increasing the prevalence of osteoporosis but British studies have not yet convincingly demonstrated any difference in smoking habits between those old women who fracture their hips and those who do not (Baker, 1980). We should also remember that there is rather more to the epidemiology of fractured femur than bone weakness, certainly the incidence of falls and the prevalence of protective responses are also important and one of the determinants of these two factors is dementia.

So far we have only considered the crude numerics of demography. A full analysis of the population changes over the next 20 years in their impact upon Health and Social Services would need to look also at the social circumstances of the future population of the aged. The importance of informal care networks in reducing the demand for statutory services has only recently been given proper attention and is still inadequately quantified. Table 6 is based on the Calderdale study in which the needs of a sample of elderly people were predicted first of all on the basis that informal support available to them would continue, and

Table 6. The Calderdale Study: Ratio of estimated provision required if all informal support withdrawn to provision required if informal support continued

	Ratio
Sheltered housing	2.1
Residential care	2.5
Psychiatric hospital	1.3
Geriatric hospital	1.6

Table 7. Relative risk of institutionalisation of ever-married women aged 75 and over by size of completed family. Whickham Urban District (1978), n = 176.

No. of children	Relative risk
0	0.9
1	1.4
2	1.3
3	0.8
4	0.5

secondly on the assumption that all such support would be removed (Howell & Parker, 1979). The ratio of the predicted provisions necessary on these two models provides some quantitative estimate of the contribution of informal care to the patterns of disability corresponding to various modalities of the Health and Social Services. Clearly the contribution is a major one.

There is abundant evidence that most of the informal support for the elderly comes from families. Most important is support from the spouse (Evans, 1981). Marriage rates and the differential in longevity of the two sexes, divorce and remarriage rates are all terms which have to be put into the equation for predicting support. The availability of children to the elderly is also clearly important. Certainly in Britain the evidence is that when children are available they will support their elderly nowadays as well as in previous times. Data from a Tyneside study comparing the family size of elderly women in institutions of various kinds compared with the size of families of women still living in the same community showed the expected inverse relationship between risk of institutionalisation and size of family over the range of 1 to 4 children (Table 7).

Interestingly, married women who had had no children showed only an average rate of institutionalisation. This raises questions about whether the absence of children had led to the formation of other supportive networks of social care or whether perhaps the relatively high institutionalisation rates of women with one or two children reflected disappointed expectations of support in old age and consequent lack of provision for alternative care. The possibility of a social class effect with the better-educated classes managing to avoid both child-bearing and institutionalisation could not, however, be excluded in our data.

Over recent years in Britain the size of families of the elderly has been declining due to the fall in birth rate during the years of economic depression between the two World Wars. Family size reached a nadir around the end of the 1930s and we can therefore expect the size of families available to the elderly to start to increase as those who produced their families after the second World War reach old age. Other factors, however, determine the availability of family help, and these will include patterns of migration and housing. Fewer old people live in the same household as their children than was the custom some years ago. There is greater mobility of the young in Britain so that children may not be as available to give care even when they are willing. We also have to recognise the fact, however unpalatable to some members of the community, that most care in practical terms comes from females rather than males. An important determinant of available care in Britain will therefore be the proportion of middle-aged women who are employed and therefore unable to give care. This proportion is greater in Britain than in other countries of the European Community and has probably been less disturbed by economic difficulties than have employment rates in men. One must emphasise that the impact of employment of women on care for the elderly is probably at least as great on the need for acute medical and health care for the elderly as on the long-term care of old people at risk of institutionalisation.

In principle these various demographic factors could be used to create a model predicting the need and demand for various types of health and social care by the elderly over the next few decades. Their effect, however, is indirect and modulated by policies and attitudes determining the conditions under which help that is potentially available is actually mobilised and given. Consideration of these factors, however, would take us well beyond the demographic context.

REFERENCES

Aniansson, A. & Gustafsson, E. (1981). Physical training in elderly men with special reference to quadriceps muscle strength and morphology. Clinical Physiology, **1**, 87-98.

Baker, M.R. (1980). The Epidemiology and Etiology of Femoral Neck Fracture. M.D. thesis, University of Newcastle upon Tyne.

Berg, S. (1980). Psychological functioning in 70- and 75- year-old people. Acta Psychiatrica Scandinavica, Suppl. 62, p. 288.

Blessed, G. & Wilson, I.D. (1982). The contemporary natural history of mental disorder in old age. British Journal of Psychiatry, **141**, 59-67.

Burnett, M. (1974). Intrinsic mutogenesis. Lancaster: Medical & Technical Press.

Collins, K.J., Doré, C., Exton-Smith, A.N., Fox, R.H., MacDonald, I.C. & Woodward, P.M. (1977). Accidental hypothermia and impaired temperature homeostasis in the elderly. British Medical Journal, **1**, 353-356.

Doll, R. (1970). The age distribution of cancer: implications for models of carcinogenesis. Journal of the Royal Statistical Society (a), **134**, 133-155.

Doll, R. & Peto, R. (1981). The Causes of Cancer. Oxford University Press.

Evans, J. Grimley (1981). Demographic implications for the planning of services in the United Kingdom. In: J. Kinnaird, J. Brotherston & J. Williamson (eds.), Provision of Care for the Elderly, pp. 8-13. Edinburgh: Churchill Livingstone.

Evans, J. Grimley (1983). Hypertension and stroke in an elderly population. Acta Medica Scandinavica, Suppl. 676, pp. 22-32.

Evans, J. Grimley (1984). Epidemiology. In: A. Martin & A.J. Camm (eds.), Heart Disease in the Elderly, pp. 9-36. London: Wiley.

Evans, J. Grimley & Caird, F.I. (1982). Epidemiology of neurological disorders in old age, In: F.I. Caird (ed.), Neurological Disorders in the Elderly, pp. 1-16. Bristol: Wright.

Evans, J. Grimley, Prudham, D. & Wandless, I. (1979). A prospective study of fractured proximal femur: factors predisposing to survival. Age and Ageing, **8**, 246-250.

Evans, J. Grimley, Wandless, I. & Prudham, D. (1980). A prospective study of fractured proximal femur: hospital differences. Public Health (London), **94**, 149-154.

Fox, R.H., Woodward, P.M., Exton-Smith, A.N., Green, M.F., Donnison, D.V. & Wicks, M.H. (1973). Body temperatures in the elderly: a national study of physiological, social and environmental conditions. British Medical Journal, **1**, 200-206.

Fries, J.F. (1980). Ageing, natural death and the compression of morbidity. New England Journal of Medicine, **303**, 130-135.

Garraway, W.M., Whisnant, J.P., Furlan, A.J., Philllips, L.H., Kurland, L.T. & O'Fallon, M. (1979). The declining incidence of stroke. New England Journal of Medicine, **300**, 449-452.

Graham, C.E., Cling, O.R. & Steiner, R.A. (1979). Reproductive senescence in female non-human primates, In: D. M. Bowden (ed.), Ageing in Non-Human Primates, pp. 183-202. New York: Van Nostrand Reinhold.

Hagnell, O., Lanke, J., Rorsman, B. & Ojes, L. (1981). Does the incidence of age psychosis decrease? A prospective longitudinal study of a complete population investigated during the 25-year period 1947-1972: the Lundby Study. Neuro Psycho Biology, **7**, 201-211.

Havlik, R.J. & Feinleib, M. (1979). Proceedings of the conference on decline in coronary heart disease. NIH Publications 79-1610. Washington DC: US Department of Health Education and Welfare.

Helderman, J.H., Vesteal, R.H., Rowe, J.W., Tobin, J.D., Andres, R. & Robertson, G.L. (1978). The response of arginine vasopressin to intravenous ethanol and hypertonic saline in man: the impact of ageing. Journal of Gerontology, **33**, 39-47.

Howell, R.G. & Parker, C.J. (1979). Assessing care requirements of elderly people in Calderdale, Reading. National Health Service Operational Group.

Isaacs, B. & Neville, Y. (1976). The needs of old people. The 'interval' as a method of measurement. British Journal of Preventive and Social Medicine, **30**, 79-85.

Johnson, M.L., di Gregorio, S. & Harris, B. (1982). Ageing, needs and nutrition: a study of voluntary and statutory collaboration in community care for elderly people. London: Policy Studies Institute.

Keyfitz, N. (1984). The population of China. Scientific American, **250**(2), 22-31.

Lewis, A. Fenton (1981). Fracture of neck of the femur: changing incidence. British Medical Journal, **283**, 1217-1220.

Marmot, M.G. (1984). Life style and national and international trends in coronary heart disease mortality. Postgraduate Medical Journal, **60**, 3-8.

Office of Population Censuses and Surveys (1973). The General Household Survey 1972. London: HMSO.

Office of Population Censuses and Surveys (1974). Morbidity Statistics from General Practice. London: HMSO.

Office of Population Censuses and Surveys (1983). Population Projections, 1981-2021. London: HMSO.

Ostfeld, A.M. (1980). A review of stroke epidemiology. Epidemiologic Reviews, **2**, 136-152.

Peacock, P.B., Riley, C.P., Lampton, T.D., Raffel, S.S. & Walker, J.S. (1972). The Birmingham Stroke Epidemiology and Rehabilitation Study, In: G.T. Stewart (ed.), Trends in Epidemiology, pp. 231-341. Springfield: Thomas.

Peto, R., Roe, F.J.L., Lee, E.N., Levy, L. & Clack, J. (1975). Cancer and ageing in mice and men. British Journal of Cancer, **32**, 411-426.

Schaie, K.W. & Strother, C.R. (1968). A cross-sequential study of age changes in cognitive behaviour. Psychological Bulletin, **70**, 671-680.

Svanborg, A. (1983). The physiology of ageing in man: diagnostic and therapeutic aspects. In: F. I. Caird & J. Grimley Evans (eds.), Advanced Geriatric Medicine 3, pp. 175-182. London: Pitman.

Taube, D.H., Winder, E.A., Ogg, C.S., Bewick, M., Cameron, J.S., Rudge, C.J. & Williams, D.G. (1983). Successful treatment of middle-aged and elderly patients with end-stage renal disease. British Medical Journal, **286**, 2018-2020.

Welin, L., Larsson, B., Svardsudd, K., Wilhelmsen, L. & Tibblin, G. (1983). Why is the incidence of ischaemic heart disease in Sweden increasing? study of men born in 1913 and 1923. Lancet **1**, 1087-1089.

Wroe, D.C.L. (1973). The elderly. Social Trends, **4**, 23-24.

CAN WE TELL OUR AGE FROM OUR BIOCHEMISTRY?

A. BAILEY
BUPA Medical Centre, Medical Information Technology,
Intercity House, Victoria Street, Bristol, U.K.

INTRODUCTION

Disease becomes more common with increasing age. Changes in connective tissue associated with the ageing process hastens degeneration of the vasculature leading to increasing cardio- and cerebro-vascular disease with age. As the internal organs function less efficiently with age, alterations in hormone levels and changes in the immune system lead to an increase in endocrine disorders, bone fragility and infectious disease in later life. Most cancers also are commoner in the elderly. The reason for this is not clear, but as life expectancy increases so does the incidence of cancer. Moreover, modern therapeutics allows many conditions to be treated well into the seventh and eighth decade to the extent that it is now often difficult to know what primary cause of death to put on the certificate of a recently decreased octogenarian.

At the BUPA Medical Centre we have been carrying out routine health screening for the last ten years. This paper presents data on biochemical constituents of blood and the changes that are seen in different age groups.

HEALTH SCREENING

The object of health screening is to detect disease at a 'sub-clinical' or a 'pre-symptomatic' phase in the belief that it will be more amenable to treatment at an early stage rather than at a late one (Beric Wright & Bailey, 1982). However, the majority of subjects attending a screening centre do not have overt disease and their characteristics can be used to study the relationship of age, among other factors, to the levels of various physiological and biochemical measurements. Four patterns of biochemical change associated with age are presented.

The biochemical measurements are estimated on serum of attenders for the screening examination, regardless of symptoms, using a continuous flow analyser. During the examination other physiological measurements are made

and a questionnaire covering personal habits, past and current medical history is completed. The record is held on computer; the data presented here are cross-sectional and based on a sample of some 10,000 males and 3,000 females.

ATHEROSCLEROSIS

The first pattern concerns the change of serum cholesterol with age. Cholesterol is the basic substance of atherosclerosis - a condition that is thought to begin with "fatty streaking" in adolescence and then progresses with increasing age as atherosclerotic plaques, consisting largely of cholesterol, become deposited on the intima of the major blood vessels eventually producing hardening and narrowing of the arteries. It is the underlying pathology of coronary and cerebral thrombosis - the major non-malignant cause of death in the Western world.

Mean serum cholesterol levels in different groups of people are directly correlated with heart attack rate. The higher the cholesterol levels, the greater the heart attack rate. Within our population the mean cholesterol levels rise with age as shown, for males, in Figure 1. Each point represents the mean total cholesterol value for one-year age groups. It will be seen that the mean level increases until the fifth decade and then flattens out, the reason for this is not completely clear - part of the effect may be due to an under-representation in the older age group, of men with high cholesterol levels who have already died of heart attacks.

The findings in women are similar but the maximum levels occur about a decade later. This is in keeping with the different incidence of heart attacks in women, which does not start to rise until after the menopause.

THE MENOPAUSE

Serum alkaline phosphatase activity varies with age (Wilding & Bailey, 1978). Mean levels are highest in children, representing bone growth activity in childhood. At maturation, levels fall substantially but then slowly increase throughout adult life. The pattern is different for the two sexes, as shown in Figures 2 and 3, the increase in males is linear but in females there is a steep increase during the menopausal years presumably reflecting liver or bone change occurring at this time. As yet, the isoenzyme of alkaline phosphatase responsible for this change (which is quite small in absolute units) has not been determined. Since it is not accompanied by any change in serum calcium levels it seems likely that the liver is the source of change.

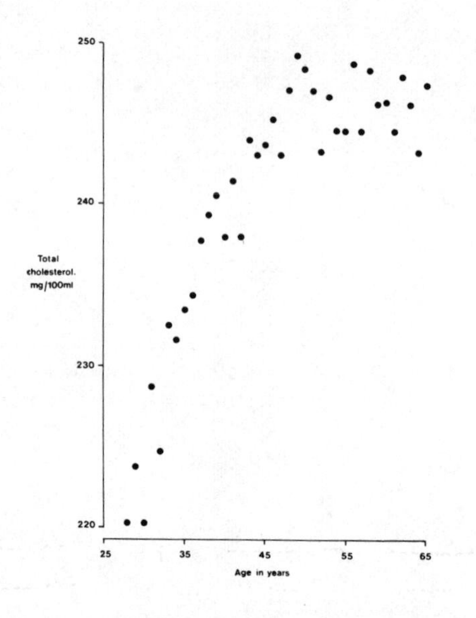

Figure 1: Serum cholesterol variation with age in males. Each point represents the mean value in each age group.

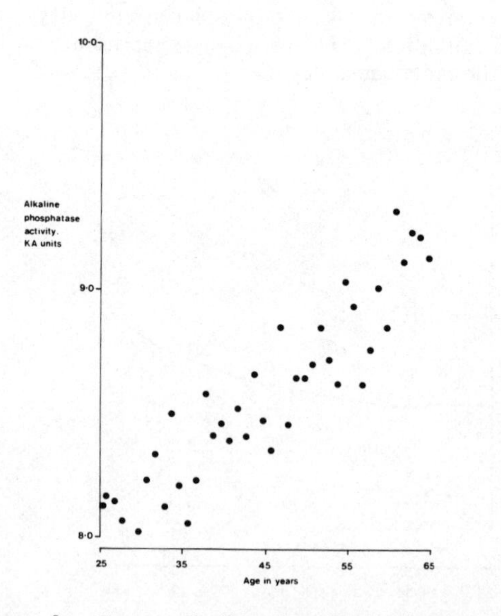

Figure 2: Serum alkaline phosphatase activity in males. Each point represents the mean value for each age group.

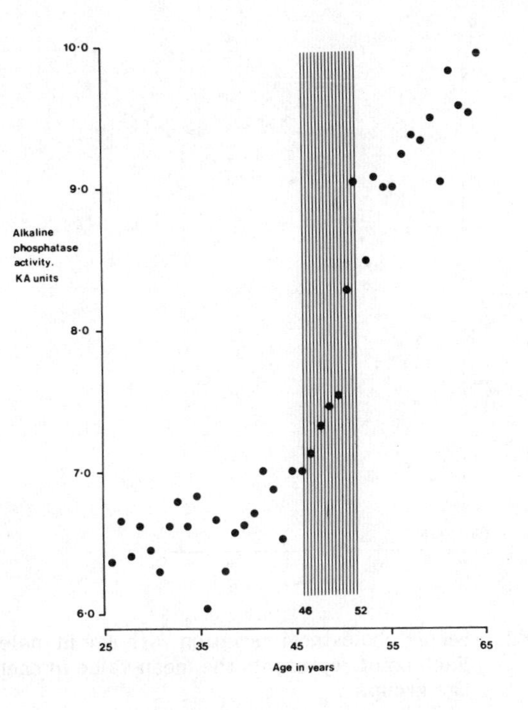

Figure 3: Mean serum alkaline phosphatase activity in females. The shaded area represents the menopause years.

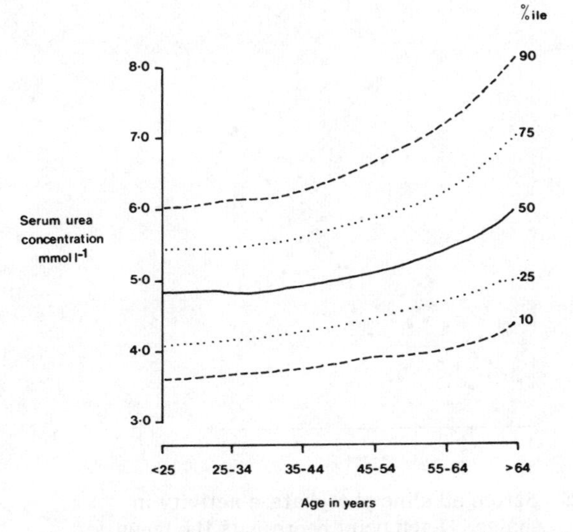

Figure 4: Percentiles for mean urea levels in males by age.

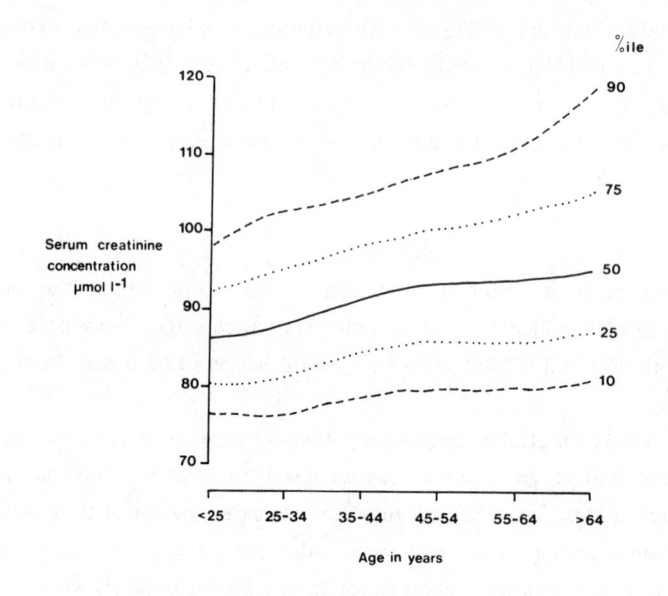

Figure 5: Percentiles for mean creatinine levels in males by age.

ORGAN FAILURE

The efficiency of various internal organs reduces with age. Hearing and vision decline, mental function deteriorates, renal function becomes less efficient. A third pattern of change associated with increasing age is seen in the biochemical constituents whose mean levels are related to the efficiency of an organ. Urea and creatinine, for example, are cleared from the body by the kidneys. The decline in renal function is reflected in the pattern of change for serum urea and creatinine (Figs. 4 and 5). Here the percentiles of mean levels are given for each decade (in males). There is a steady rise with age. This pattern is also interesting because of the wide variation in mean values for each age group - a 20% variation across the percentiles. This is commented on later.

ORGAN ADAPTATION

In health, organ function responds to external stimuli to maintain internal homeostasis. In some cases this is automatic, for example dealing with a fatty meal. In others, enzyme systems need to be induced. The response to alcohol ingestion is an example. Alcohol is mostly metabolised by alcohol dehydro-

genase - but if this system becomes overloaded, other systems may be induced. Gamma glutamyl transferase (GGT) activity reflects the microenzymal oxidation system (MEOS) for oxidising alcohol (Robinson et al, 1979). Figure 6 shows how mean levels of GGT activity increase in the 35-44 age group and thereafter. This is probably due to hepatic adaptation to an adult consumption of alcohol metabolised via MEOS.

DISCUSSION

Can these patterns of biochemical change help to determine the pathophysiological age of a subject? Not as precisely as, say, the X-ray of a wrist will give the age of a child. But there is a useful lesson to be learnt from each pattern.

Recent American studies have shown that by reducing serum cholesterol levels using diet and drugs a diminution in heart attacks is achieved (Lipid Research Clinics, 1984). This has provided the strongest evidence to date for a causal link between serum cholesterol level and heart attack. Further studies on lipoproteins, which transport cholesterol in the blood, have shown that the ratio of high density lipoprotein to low density lipoprotein is a powerful predictor of cardio-vascular disease (Williams et al, 1979). Using this and other known risk factors, it is possible to estimate a 'pathological' age in apparently healthy individuals.

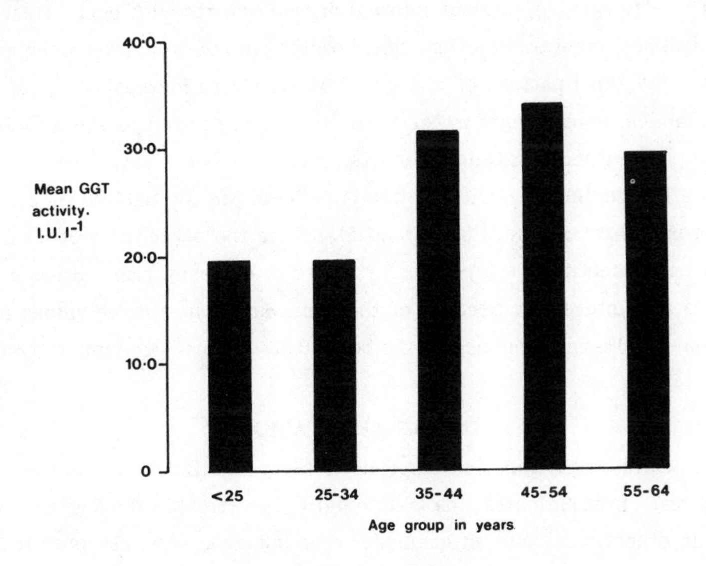

Figure 6: Mean GGT activity in males by decades.

Biochemical modifications associated with the menopause may be useful in two ways. First, they may predict fecundity - and thus allow the completion of contraceptive measures at the appropriate time even if this is not obvious clinically. Secondly, certain patterns may be associated with peri-menopausal symptoms thus helping to identify the appropriate treatment. Alkaline phosphatase activity is given here as one example of the biochemical parameters that change at the time of menopause; there are many others.

Measurements of organ failure may be useful in elderly patients who are being treated with a number of drugs. Drugs are metabolised in the liver and excreted by the kidney. Much of the individual variation in therapeutic response and idiosyncratic reaction to drugs may be explained by the efficiency of the metabolising and excretion pathways. The variation demonstrated in Figure 4, mean urea levels in men, provides a good example of this phenomenon.

Similarly, individual variation in response to alcohol may be related to the body's method of metabolising it, and this is reflected in enzyme activity. GGT levels can be used to predict some of the physical complications of alcohol abuse. In our follow-up studies (unpublished), the death rate in heavy drinkers with high GGT levels is almost twice that of heavy drinkers with low GGT levels.

Biochemical measurements made in health may be used to predict the onset of disease occurring in the future. As we gain prospective experience we may be able additionally to throw light on the aetiology of the ageing process.

REFERENCES

Beric Wright, H. & Bailey, A. (1982). Screening well populations. In: R.S. Schilling (ed.), Occupational Health, 2nd edn., p. 203. London: Butterworth.

Lipid Research Clinics Coronary Primary Preventive Trial Results (1984). Journal of the American Medical Association, **251**, 351-364.

Robinson, D., Monk, C. & Bailey, A. (1979). The relationship between serum gamma-glutamyl transpeptidase level and reported alcohol consumption in healthy men. Journal of Studies on Alcohol, **40**, 896-901.

Wilding, P. & Bailey, A. (1978). The normal range. In: V. Marks (ed.), Scientific Foundations of Clinical Biochemistry, p. 451. London: Heinemann.

Williams, P., Robinson, D. & Bailey, A. (1979). High-density lipoprotein and coronary risk factors in normal men. Lancet **i**, 72-75.

DIETARY MANIPULATION OF AGEING: AN ANIMAL MODEL

B. J. MERRY
The Wolfson Institute, The University of Hull, Hull, U.K.

INTRODUCTION

In addition to studies in man, the scientific necessity of using animal models has long been recognised for they allow an insight into disease states and physiological processes which otherwise would not be possible. With regard to the difficult question of extrapolation of data from animals to man, it seems realistic to surmise that from a similar biochemical background such projections are valid. The widespread use of animal models to investigate the aetiology of numerous clinical conditions supports this contention (Andrews et al, 1979). One of the main goals of experimental ageing research is to understand the mechanisms of ageing in order to rationally approach the health problems of the aged as well as to increase the quality of life in old age (Hollander, 1979). Within this context research interest is currently focussed on exploring the potential of a unique animal model in order to understand more fully the mechanisms of ageing and the biochemical aetiology of chronic age-related pathologies. Advantage has been taken of the observation, so far restricted to rodent species within the mammals, that manipulation of diet composition or controlled feeding will increase maximum lifespan and alter the pattern of degenerative diseases (Moment, 1982).

DIETARY RESTRICTION MODEL

The extension of lifespan in rats by dietary means was first reported in 1917 (Osborne et al, 1917) to be further investigated by McCay and his colleagues in the 1930s (McCay et al, 1935), Berg and Simms in the 1960s (Berg & Simms, 1960, 1961), to be followed by the extensive studies of Ross (1959, 1964, 1966, 1969, 1972). Subsequent to the slow historical development of this model system, research interest is now widespread in utilising diet as a probe to investigate the ageing process. An extension of the lifespan has been observed in rodents in spite of widely differing experimental designs, species and strain of

animal, composition and caloric value of the diet. This aspect has been comprehensively reviewed in Holehan (1984).

One of the simplest and most effective designs has been to retard growth by limiting access to the normal diet such that the body weight of experimental animals is maintained at 50% that of the age-matched *ad libitum* fed rats (Masoro et al, 1980; Merry & Holehan, 1979, 1981). The restricted feeding regime is imposed either at weaning or in the immediate postweaning period with the animals being individually housed to prevent competition for the limited food. The age at increase in the rate of mortality is delayed and an extension of 40-66% in maximum lifespan is observed (Fig. 1). Controlled underfeeding will prolong lifespan in both sexes of rats, and the effect has been recorded for rats, mice and hamsters (Stuchliková et al, 1975).

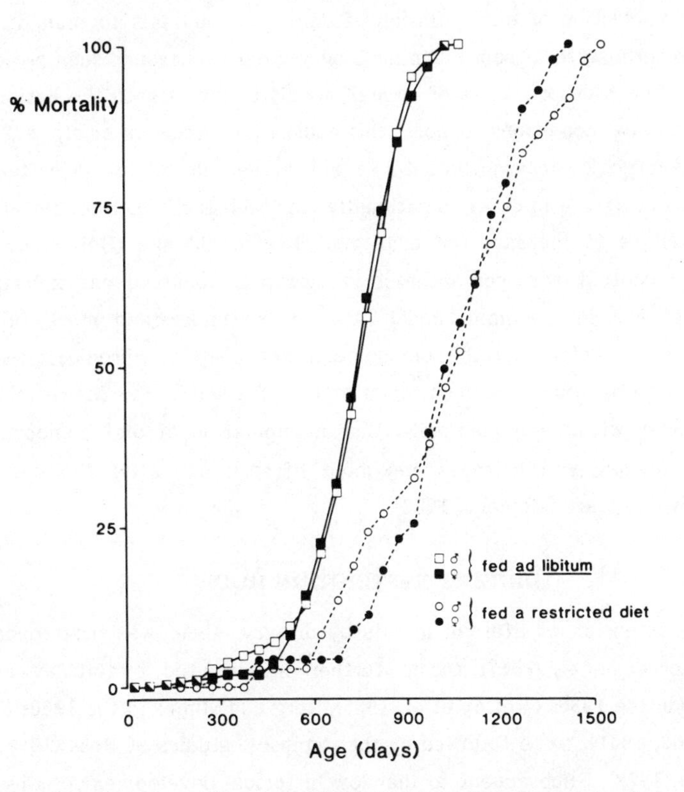

Figure 1. Mortality profiles of CFY Sprague Dawley rats fed *ad libitum* (□♂ ■♀) or a restricted diet (○♂ ●♀) such that body weight was maintained at 50 per cent that of the fully fed animals. Group numbers: *ad libitum* = 200, 397; dietary restricted = 100, 100. (Data from Merry & Phillips, 1981.)

Confirmation of the reproducibility of the effect of dietary restriction even across differing strains is seen in the data from Yu et al (1982) and Merry and Holehan (1981, 1985a) who reported a 49% extension of maximum lifespan for Fischer 344 rats and a 42% extension in CFY Sprague Dawley male rats. The time to double the rate of mortality (T_d) for *ad libitum* fed rats was 102 days (Merry & Holehan, 1985b), 99 days (Ross, 1959) and 104 days (Yu et al, 1982), while restricted feeding increased the T_d to 203, 210 and 189 days respectively.

Ross (1959) using over 1600 Charles River COBS male rats investigated the effects of different dietary components on longevity. The rats were distributed into five groups, one fed a commercial diet *ad libitum* and the other four calorically restricted diets consisting of low protein-low carbohydrate (LP-LC), low protein-high carbohydrate (LP-HC), high protein-low carbohydrate (HP-LC), and high protein-high carbohydrate (HP-HC). Adult body weight was an increasing linear function of cumulative calorie intake and in all cases the restricted rats had a greater life expectancy than those fed *ad libitum*. The high protein intake groups were markedly superior in the first half of life and the HP-LC group had the greatest average survival. The HP-HC group had a steep late mortality as did the LP-HC group. The LP-LC group, despite a high early mortality from caecal impaction, showed the best late life survival. However it must be emphasised that the diet designated low protein-low carbohydrate was 21.62% casein and 54.05% sucrose and these terms therefore are only relative. The caloric intakes of these five dietary groups are linearly related to their lifespan (Ross, 1969) with a slope of about -3.9 days/kcal/day. The lifetime energy intake per gramme of tissue only varies by 7.5% over the five groups and Sacher (1977), in reviewing life-prolonging experimental procedures, concluded that the anti-ageing action of food restriction is due to a reduction in the rate of metabolism per unit of body mass. This was based on the idea of a constant lifetime caloric consumption per gramme of body tissue and that food restriction prolongs life by delaying the time required to reach the total. Re-examination of the food intake data of Masoro et al (1982), Merry and Holehan (1979) and Holehan (1984) does not support this concept. In these investigations the diet-restricted rats consumed a greater number of calories per gramme body mass during their lifetime than did the rats fed *ad libitum*.

Throughout the numerous reports utilising dietary restriction to increase lifespan - underfeeding, caloric restriction, etc. - it has been unclear whether the beneficial effects were due to caloric or to protein restriction. Divergent

results have been reported (Miller & Payne, 1968; Ross & Bras, 1973, 1975; Nakagawa et al, 1974; Leto et al, 1976; Beauchene et al, 1979; Payne, 1979) originating from the diverse experimental approaches adopted. Davies et al (1983) have reinvestigated this confused aspect of the model by providing caloric restricted rats with the same amount of protein as their paired *ad libitum* fed controls but one-third fewer calories. Rats fed the calorically restricted diets (67% of *ad libitum* fed controls) survived longer than those fed *ad libitum* even when both caloric groups consumed the same amount of protein. Although caloric intake had a greater effect on survival than protein intake in this study, low levels of dietary protein in both caloric restricted and *ad libitum* fed groups were associated with decreased survival rates. However, energy intake determines the amount of protein available for protein synthesis; as the severity of energy restriction is increased more protein is utilised for energy purposes. Thus an experimental regimen which involves restriction of food supply inevitably introduces an element of both energy and protein deficiency and there is no way in which the effects of protein *per se* can be totally separated from that of energy (Payne, 1979).

TIMING OF INITIATION OF FOOD RESTRICTION

A number of laboratories have addressed the question as to whether there are critical periods in the life of an animal during which underfeeding is most effective in extending lifespan. When rats were dietary restricted for seven weeks postweaning and then fed *ad libitum* throughout life there was a 20% reduction in mortality risk (Ross, 1972), but no extension of maximum lifespan. Reversal of this schedule increased the LD_{50} and maximum lifespan but not to the same extent as dietary restriction throughout postweaning life. Ross (1966) and Ross and Bras (1974) utilised another approach, feeding rats *ad libitum* from 21 days to 300 days at which age they were divided into groups whose diet varied in protein content from 8 to 50.9% and whose level of allotment of a particular diet ranged from 40 to 100% of the amount consumed by the *ad libitum* fed animals. A mortality index for each group was calculated from their age-specific mortality rates after 300 days. Animals offered 60–70% of *ad libitum* intakes of the commercial diet (23% protein) or 52% of *ad libitum* intake of a diet which contained 21.6% vitamin free casein, showed beneficial effects on their mortality indices. This finding was confirmed by Beauchene et al (1979) in which *ad libitum* fed male rats had a mean lifespan of 133 weeks compared to 163 weeks for animals dietary restricted throughout postweaning life. A

schedule of feeding *ad libitum* for one year followed by dietary restriction or feeding *ad libitum* at one year after dietary restriction, extended mean lifespan (150 and 149 weeks respectively), although the effect was less than in animals with dietary restriction throughout life. Stuchliková et al (1975) in a study of similar experimental design concluded that food restriction limited to the first year of life was the most effective nutritional regimen in prolonging life.

Weindruch et al (1979) using two strains of long-lived mice instigated a gradual restriction of food intake (72% of *ad libitum* intake for one month then 56% thereafter) at 12 months of age and increased mean survival time by 10-20%. The effect of changing the diet to one containing only 4% protein, on animals which had previously been given a stock diet *ad libitum* until skeletal growth has ceased, was reported by Miller and Payne (1968). This treatment resulted in an increased average lifespan of the same order as that induced by lifelong restriction.

This issue of the timing of underfeeding with respect to maximising lifespan has still not been resolved. The effect on lifespan of returning previously dietary restricted animals to full-feeding at varying times throughout their life has been more controversial. Nolen (1972) concluded that restriction of dietary intake either throughout life or from 12 weeks after weaning until death prolonged lifespan, whereas early diet restriction followed by *ad libitum* feeding did not. At 13 and 19 months of age, Barrows and Roeder (1965) restricted by 50% the intake of a commercial diet containing 25% protein. This failed to increase lifespan and the mean age at death for all restricted rats was slight (1.8 months) but significantly lower than that of the *ad libitum* fed controls. Re-investigation of this issue by Merry and Holehan (1985b) who returned male rats to *ad libitum* feeding after 7, 39, 159, 236 and 344 days of restricted feeding, concluded that approximately one year of restricted feeding is obligatory in CFY rats before a significant (10%) extension in maximum lifespan can be observed.

Although it may be said that treatments which are continued for a longer fraction of postweaning life generally have a greater effect on extending lifespan, it has been clearly demonstrated that the immediate postweaning period is not the only phase of the lifespan sensitive to the effects of underfeeding. Many of the investigations reported quote only mean lifespan and the effects of restricted feeding at more advanced ages may reflect only an altered pathology profile and may not be retarding the rate of ageing.

DIETARY RESTRICTION AND AGE-RELATED PATHOLOGY

That the age-related pathologies of rodents are exquisitively sensitive to the dietary history of the animals has been demonstrated by a number of studies. The early work of McCay et al (1943), Tannenbaum and Silverstone (1949, 1953), and later Berg and Simms (1960, 1961), Ross (1959, 1964), Ross and Bras (1965, 1973), Ross et al (1970), Tucker et al (1976) and Yu et al (1982) have confirmed the increased resistance to disease and delay in the appearance of age-related pathologies resulting from dietary restriction. There is not only a delay in the onset of chronic glomerulonephritis and of cardiac and vascular lesions but also a reduction in the cumulative lifetime incidence of these lesions. Myocardial fibrosis, peribranchial lymphocytosis, periarteritis, prostatitis and endocrine hyperplasias are decreased by 50-90% in dietary restricted animals and the cases which do develop are usually mild and seen at very late ages (Ross, 1959, 1964; Berg & Simms, 1960).

Most of the emphasis relating chronic underfeeding to pathology has concentrated on the incidence of neoplastic lesions. A comprehensive study of diet-related changes in tumour incidence has been presented by Ross and Bras (1965, 1973) and Ross et al (1970). They reported that the incidence of neoplasms was a linearly increasing function of caloric intake level or body weight. There was a reduction in the frequency of tumours of the pituitary gland, lung and pancreatic islet cells but no significant alteration in the incidence of tumours of the thyroid gland, urinary bladder and tumours of soft tissue origin, although there was a delay in the age of onset of these tumours. Conversely a significant increase was observed in the incidence of adrenal and parathyroid gland tumours and of reticulum cell sarcomas of the lymphoid organs. Malignant epithelial tumours were also more prevalent in dietary restricted rats (Ross & Bras, 1971). Certain forms of nutritional intervention limited to early life followed by an *ad libitum* feeding regimen which is normally conducive to a high incidence of neoplasia, substantially reduces tumour incidence in later life (Ross & Bras, 1971). Payne (1979) has commented that in this area there seems to be unequivocal evidence of a specific effect of protein intake. However, the effect is complex, in that some types of tumours are commonest with high protein intake and others with low protein intake.

A recent study has investigated one of the possible molecular mechanisms of reduced carcinogenesis and delayed ageing in dietary restricted rats, (Woodhead et al, 1985). O^6-Methylguanine produced in DNA by simple alkylating carcinogens is considered to be the primary mutagenic lesion and is

implicated in carcinogenesis induced by these agents. Undoubtedly, the ability of a tissue to remove lesions from DNA is only one of the important factors that affect its susceptibility to cancer formation. However, DNA alkylation in rodent tissues is apparently uniform immediately after exposure and therefore target size and relative repair rates play major roles in determining the subsequent distribution of tumours (Swann & Magee, 1968). Striking and consistent differences in liver, spleen, brain and kidney acceptor protein levels were recorded which correspond to the sensitivity of these tissues to tumour formation after *in vivo* exposure to alkylating agents. However no consistent differences between levels of acceptor protein in any tissue could be correlated with underfeeding.

Caloric restriction during growth in mice and rats has been shown to delay the development of the immune system but also to prolong the maintenance of immunologic competence (Weindruch et al, 1979). Underfeeding from weaning reduces thymus growth, alters the timing of thymic involution (Weindruch et al, 1979) and retards the age-associated autoimmunity to cerebral tissue observed in C57BL/6 (Nandy, 1982). It is of interest that restricted feeding delays physiological ageing and reduced tumour onset and development in specified pathogen-free (SPF) and germ-free rodents (Yu et al, 1982; Pollard M., personal communication).

With regard to the long-term modification of age-related pathology the report of Lloyd (1984) on the WKY/SHR (spontaneously hypertensive rat) is noteworthy. The development and maintenance of hypertension in the SH rat closely resembles the course of essential hypertension in humans in that (1) the developing phase in the SH rat occurs before the animal is mature whereupon a secondary, slower increase in blood pressure continues throughout life, (2) the lifespan of the SH rat is shorter by about one third when compared to its WKY normotensive progenitor and (3) the cause of death in the SH rat has generally been ascribed to end-organ damage in the heart, adrenal, kidney and brain, consequent to the elevated blood pressure (Lloyd & Boyd, 1981). When SH rats were maintained on restricted feeding (40% of a typical laboratory rat diet), lifespan was increased and was similar to that of the normotensive WKY dietary restricted rat (Fig. 2). The mean lifespan of food restricted WKY rats increased from 24 to 32 months whereas the mean lifespan of the SH rat under similar dietary restricted conditions increased from 18 to over 30 months. The increased lifespan was associated with a significant decrease in the incidence of six specific lesions typical of the SH rat; interstitial fibrosis of the kidney,

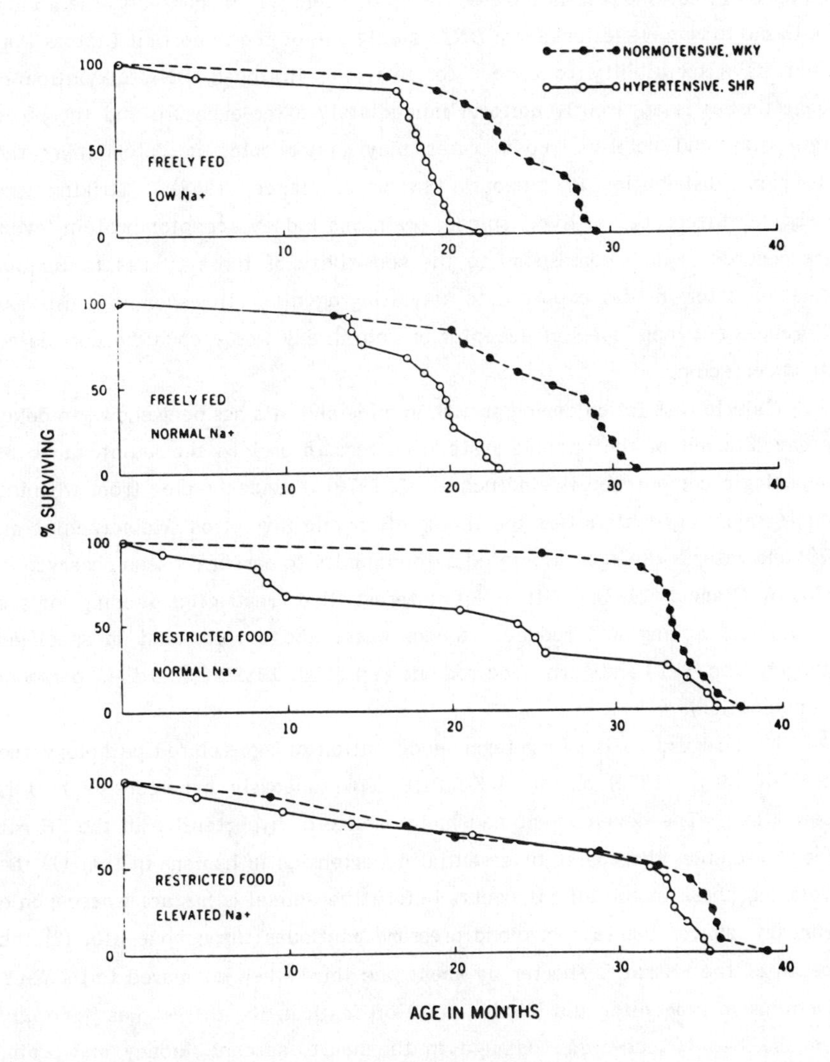

Figure 2. Survival curves of fully-fed and dietary restricted spontaneously hypertensive rats and WKY normotensive controls. N = 12 SHR and 12 WKY in each group. (Data redrawn from Lloyd, 1984.)

myocardial oedema, fatty infiltration of the myocardium, atrial thrombosis, congestion and vacuolisation in the adrenal. The frequency of these lesions in underfed SH rats was not significantly different from that recorded in dietary restricted WKY normotensive rats (Lloyd, 1984). The study of Lloyd and Boyd

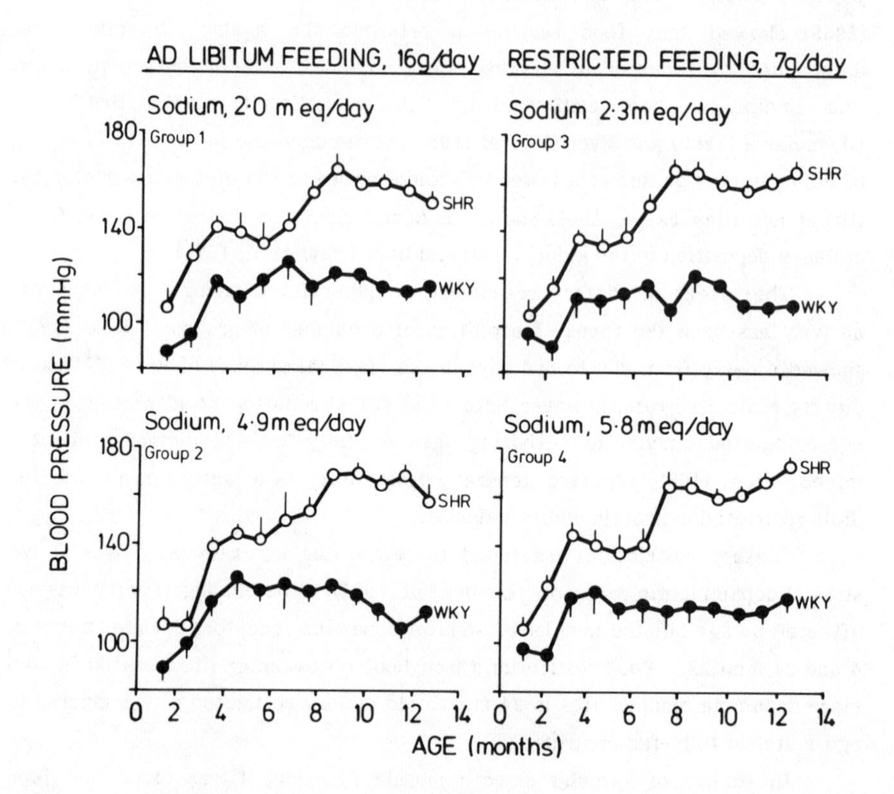

Figure 3. The effect of dietary restriction and dietary sodium on the development and maintenance of hypertension in the spontaneously hypertensive rat. (Data redrawn from Lloyd & Boyd, 1981.)

(1981) indicated that long-term caloric restriction does not mitigate the development of hypertension in the SH rat but it does allow a genetically hypertensive animal to reach a normal lifespan by protecting against end-organ damage resulting from hypertension (Lloyd, 1984) (Fig. 3).

DIETARY RESTRICTION AND PARAMETERS OF AGEING

It is now the consensus that dietary restriction acts at a fundamental level in the cell biochemistry, for the extension of lifespan is far greater than can be explained by the elimination or postponement of the diseases characteristic of the ageing rodent. A study of a number of physiological and biochemical variables demonstrating significant age-related changes suggests that animals maintained on restricted feeding are retained in a physiologically

younger condition than age-matched fully-fed control rats. Chvapil and Hrůza (1959) showed that food restriction retarded the ageing of collagen as determined by tail-tendon breaking time, an index of collagen cross-linkage. This finding has been confirmed by Giles and Everitt (1967), Hrůza and Hlaváčková (1969) and Everitt et al (1981). The decrease in the rate of ageing of collagen may be due to a lower tail temperature in the dietary restricted rat (Hrůza and Hlaváčková, 1969) and it has been shown also that there is a delay in collagen deposition in the kidney, lung and liver (Deyl et al, 1971).

The effect of dietary restriction on age-related changes in enzymatic activity has been the focus of attention of a number of groups. Ross (1959) showed a delay from 200 to 600 days in the maximal activity of liver ATPase in dietary restricted rats, however Leto et al (1976) reported no alteration in the age-associated enzymatic activities between fully-fed and dietary restricted mice. Ross (1969) reported generally lower enzymatic activities in rats fed diets restricted in protein and/or calories.

Dietary restriction was found to retard the age-associated loss of rat striatal dopaminergic receptors (Levin et al, 1981). The binding affinity was not affected by age but the number of striatal dopamine receptors declined between 4 and 24 months. Food restriction throughout postweaning life resulted in 50% more dopamine binding sites in 24 month old dietary restricted rats compared to age-matched fully-fed controls.

In studies of vascular smooth muscle function, it was found that food restriction prevented the age-related loss in tension development that occurred late in life in *ad libitum* fed rats (Herlihy & Yu, 1980) but in the gastrocnemius muscle the only parameters affected by food restriction were a delay in the age-related decline in mass and associated increase in collagen (Yu et al, 1982).

The adipose mass of diet-restricted rats is smaller as a result of fewer and smaller adipocytes (Bertrand et al, 1977, 1980; Masoro et al, 1979). The responsiveness of adipocytes to glucagon and epinephrine declines with age with the loss of responsiveness to glucagon preceding the loss of responsiveness to epinephrine (Yu et al, 1980). This profile of loss of responsiveness of adipocytes to lipolytic hormones with age is delayed by dietary restriction (Bertrand, 1983; Cooper et al, 1977). It was also found by Masoro et al (1979, 1980) and Liepa et al (1980) that post-absorptive serum triglyceride concentrations were lower in food-restricted rats than age-matched fully-fed animals and that the age-related increase in serum cholestrol and phospholipid concentrations were less severe in underfed animals.

The age-related accumulation of tissue lipofuscin has been used as a biomarker of ageing. Enesco and Kruk (1981) found that brain and cardiac lipofuscin was lower when male mice were fed low protein diets than when fed high protein diets *ad libitum*.

DIETARY RESTRICTION AND REPRODUCTIVE DECLINE

One of the characteristics of ageing in the female rat is the gradual onset of infertility as shown by a lengthening of the oestrous cycle and decrease in litter size. Unlike many of the reported parameters of ageing, the decline of which have been shown to be delayed by dietary restriction, the extension of fertility beyond the normal age in the female is of interest because it demonstrates the functional retention of a fully integrated system involving several tissues and organ systems. Merry and Holehan (1979) reported that chronic dietary restriction (continuous adjustment of food intake so that body weight was 50% of the fully-fed control rats), not only extended lifespan, but retarded the onset of puberty and extended reproductive lifespan. Normal vaginal cyclicity persisted for more than 18 months, as compared to less than 12 months for control rats. It is clear that the observations on vaginal cytology indicate a slowing down in the rate of reproductive ageing in the dietary restricted rats. In underfed animals not only was there a retention of regular cycles to a greater age but there was no change in the length of the cycle at the ages tested (Merry & Holehan, 1979; Holehan, 1984). There was a later onset in irregularities of the oestrous cycle in restricted rats, no anoestrus vaginal smears being found until 730-750 days and at this age 61% of the group were still cycling normally and only 8% exhibited anoestrus smears.

Earlier work with mice on restricted feeding has indicated either a state of near sterility when caloric intake was restricted from weaning (Ball, et al, 1947) or total sterility when the restriction was imposed on the adult animal (Visscher et al, 1952). This latter group had shown that mice dietary restricted from weaning could produce litters but were unable to wean them. From the data of Merry and Holehan (1979), Merry et al (1984), and Holehan (1984) it is clear that rats retained on a restricted diet during the course of their pregnancy were able to produce multiple live litters to an age far exceeding that of the fully-fed females (Table 1). Control CFY female rats showed a rapid decline in fertility from 270 days and were infertile by 500 days, while in the dietary restricted female maximum fertility was achieved about 500 days with some virgin females breeding and rearing litters as late as 900 days.

Table 1. Effect of chronic dietary restriction on breeding performance in rats of different ages. (Data from Merry & Holehan, 1979.)

Age (days)	No. of rats	% Pregnant	Litter size (mean, s.e.m.)	Weight of young at weaning (mean, s.e.m.)	No. weaned/ no. liveborn
Control rats					
150	45	100	12.5 0.4	43.1 1.1	0.88
210	20	100	8.3 0.5*	42.3 1.5	1.00
270	20	75†	6.3 0.5*	40.1 2.8	0.68
420	30	20†	1.0 0.4*	-	0.00
450	28††	0	-	-	-
Experimental rats					
300	60	13	6.0 0.7	39.0 0.8	0.67
330	50	30†	10.0 0.2	38.8 1.5	0.68
510	50	80†	5.0 0.5	36.5 0.6	0.89
840-930	32	25†	4.5 0.8	35.0 0.6	0.86

* Significantly different from value at 150 days, $P < 0.01$ (Student-Newman-Keuls multiple range test following a nested analysis of variance).

† Significant nonhomogeneity of percentages, $P < 0.01$ (method of Brandt & Snedecor).

†† Two animals died before the breeding study could be undertaken.

As the irregularities in oestrous cyclicity have been attributed to changes in ovarian endocrine secretion and reduced capacity of the hypothalamo-pituitary system to respond to gonadal hormone feedback (Huang et al, 1978; Miller & Riegle, 1979; Lu, 1983), it would appear that these factors remain undisturbed in restricted female rats for a longer period. This is in agreement with the hormonal profile reported for dietary restricted female rats by Holehan (1984). The data reported on the hormonal profiles of the oestrous cycle and vaginal cytology with age go some way to explain the reduced incidence of congenital malformations in the offspring born to underfed females (Merry et al, 1985; Holehan, 1984). The incidence of birth defects was recorded in over 4000 pups born to fully-fed and dietary restricted female rats. While the incidence of malformed young born to fully-fed controls was approximately 1%, no malformations were observed in pups born to dietary restricted mothers.

In both humans and rats, preovulatory ageing of oocytes is detrimental to development (Mikamo, 1970). Old rats with extended oestrous cycles appear to have an increase in developmental abnormalities when compared to younger

controls. Fugo and Butcher (1971) have shown that an extension of the oestrous cycle to six days produces a marked increase in the number of abnormal and unfertilised one-cell ova, degenerating embryos and degenerating conceptuses. It is possible in young adult rats to induce an experimental delay of 48 hours in the timing of ovulation by injecting sodium pentobarbital to delay ovulation (Fugo & Butcher, 1966). This delay in ovulation results in follicular ageing associated with developmental defects, a decrease in the implantation and fertilisation rate linked with a three-fold increase in polyspermy.

Associated with extended oestrous cycles in aged fully-fed rats is an early rise in the plasma levels of oestradiol-17β in relation to the timing of ovulation. Binding of the endogenous oestrogen with a specific antiserum during this period of the cycle prevented the increase in developmental abnormalities (Butcher & Pope, 1979) but the incidence in developmental defects resulting from delayed ovulation was restored by administration of diethylstilboestrol, an oestrogenic compound not bound by the antiserum (Table 2). The relative early rise of oestrogen in the oestrous cycles of the ageing fully-fed rat has the consequence that the preovulatory follicle is contained for a prolonged time in an environment of elevated oestrogen which may result in disruption of the oocyte RNA stacking arrangement and subsequent abnormal development (Butcher & Page, 1981; Peluso et al, 1980).

Table 2. Effects of neutralisation of oestrogen early in the oestrous cycle on embryonic development by midgestation in rats with delayed ovulation. (Data redrawn from Butcher & Pope, 1979.)

			% Surviving embryos		
Groups[a]	Implantation rate (%)	Postimplantation death (%)	Normal	Abnormal	Retarded
Control	85	5	92	1	7
PB[b]	43	22	57	9	34
ASE[c]+PB	89	8	81	1	18
DES[d]+ASE+PB	75	18	70	7	22

[a]n = 27-29 litters/group at day 11 of gestation.
PB[b] = sodium pentobarbital delay of ovulation for 48 hours.
ASE[c] = antisera to oestradiol on days 1 and 2 of cycle.
DES[d] = diethylstilboestrol (0.25 μg day 1, and 0.5 μg day 2).

The hormonal profiles of rats on a restricted diet show no significant difference in the overall pattern of oestradiol-17β across the oestrous cycle but the total amount of the hormone released in dietary restricted females is 46% that of the control animals. Higher levels of follicle stimulating hormone (FSH) are observed associated with an early release of the preovulatory peak of luteinising hormone (LH) in underfed females. As Holehan (1984) points out, such an alteration in the hormonal profile has, as one consequence, a reduced exposure of the follicular ovum to high levels of oestradiol-17β. In the context of the experimental manipulation of Fugo and Butcher (1971) on the effect of oestradiol-17β on congenital malformations, this observation is clearly of some importance. Berg (1965) restricted the food intake of 10-11 weeks old Sprague-Dawley rats commencing on the day of mating and found that in both fully-fed and restricted groups (75%, 50% and 25% of *ad libitum* diet), no fetal malformations were seen. It is therefore feasible that the oestrous cycle of dietary restricted female rats reflects a more precise synchrony between the priming of follicles by FSH, the increase in oestrogen and the release of LH to induce ovulation.

THE OPERATIVE EFFECT OF RESTRICTED FEEDING

The mechanism or operative effect of dietary restriction is unknown but it may well be a complex central neuroendocrine response mediated by the hypothalamus as suggested from the maintenance of normal oestrous cycles and fertility in these animals. It is possible to mimic certain aspects of this model by hypophysectomy of rats at 70 days of age coupled with replacement cortisone therapy (for a review of this model see Everitt, 1982). Hypophysectomised rats show a reduced incidence of pathological lesions, retarded collagen cross-linkage, depressed heart-rate, haemoglobin, white cell count and creatinine excretion. However the interpretation of the data from underfeeding studies as indicative of a dietary hypophysectomy (Samuels, 1946; Everitt, 1982; Comfort, 1979; Moment, 1982) is clearly too simplistic in view of the extended fertility recorded in these animals. It is still unclear how central are the endocrine changes which result from chronic underfeeding, to the reduction in protein and DNA synthesis as observed in the food restricted rat (Merry & Holehan, 1985a), and to the delayed maturation of the immune system (Walford et al, 1974). The hypophysectomised male Wistar rat model of Everitt may well represent an alternative means of chronic dietary restriction with the hormonal changes being secondary to the reduced food intake.

Barrows and Kokkonen (1982) have discussed the possibility that a reduction in food intake which results in decreased protein synthesis may represent decreased use of the genetic code. They argue that the rate of occurrence of the programmed events encoded in the genome which are characteristic of development are influenced or controlled by the rate of synthesis of specific regulatory RNAs and proteins. Thus it is proposed that a way to retard the rate of ageing would be to decrease the rates of synthesis of specific RNAs and proteins during the life of the organism. One of the persistent changes in the tissues of chronically underfed rats is the reduction of RNA (Frolkis, 1982; Merry & Holehan, 1985a). This is associated with a decrease in the immediate postweaning rate of *in vivo* DNA synthesis (Merry & Holehan, 1985a) and transformation of hepatocytes to a higher ploidy state (Enesco & Samborsky, 1983; Gahan et al, 1985). However it has been shown that prolonged underfeeding results in increased transcription of rRNA coupled with an increased turnover of rRNA and ribosomes (Kawada et al, 1977), and increased synthesis and degradation of tRNA and the 4-18S component of the insoluble RNA fraction (Srivastava et al, 1978).

The use of the inhibitor of transcription, olivomycin, has been shown to increase the maximum lifespan of rats by 25-30% (1481 ± 9 days compared to 1153 ± 11 days), an effect comparable with dietary restriction (Frolkis et al, 1976). There are two interesting aspects of this report. Firstly the maximum lifespan in a mammal can be significantly increased by pharmacological means of known site of action, and secondly the administration of olivomycin was not commenced until the rats were 600 days.

Clearly further studies are needed to understand the mechanisms of action of dietary restriction, but as an experimental probe for understanding how the modification of gene expression during development can subsequently affect reproductive and somatic lifespan, age-related pathology and the expression of hypertension-related end-organ damage, it promises to be a powerful tool.

REFERENCES

Andrews, E.J., Ward, B.C. & Altman, N.H. (1979). Spontaneous Animal Models of Human Disease, vol. 1. London: Academic Press.

Ball, Z.B., Barnes, R.H. & Visscher, M.B. (1947). The effects of dietary caloric restriction on maturity and senescence, with particular reference to fertility and longevity. American Journal of Physiology, **150**, 511-519.

Barrows, C.H. & Kokkonen, G.C. (1982). Dietary restriction and life extension - biological mechanisms. In: G.B. Moment (ed.), Nutritional Approaches

to Aging Research, pp. 219-243. Florida: CRC Press Inc.

Barrows, C.H. & Roeder, L.M. (1965). The effect of reduced dietary intake on enzymatic activities and lifespan in rats. Journal of Gerontology, **20**, 69-71.

Beauchene, R.E., Bales, C.W., Smith, C.A., Tucker, S.M. & Mason, R.L. (1979). The effect of food restriction on body composition and longevity of rats. Physiologist, **22**, 8.

Berg, B.N. (1965). Dietary restriction and reproduction in the rat. Journal of Nutrition, **87**, 344-348.

Berg, B.N. & Simms, H.S. (1960). Nutrition and longevity in the rat. II. Longevity and onset of disease with different levels of food intake. Journal of Nutrition, **71**, 255-263.

Berg, B.N. & Simms, H.S. (1961). Nutrition and longevity in the rat. III. Food restriction beyond 800 days. Journal of Nutrition, **74**, 23-32.

Bertrand, H.A. (1983). Nutrition-aging interactions: life-prolonging action of food restriction. Review of Biological Research in Aging, **1**, 359-378.

Bertrand, H.A., Masoro, E.J. & Yu, B.P. (1977). Postweaning food restriction reduces adipose cellularity. Nature, **266**, 62-63.

Bertrand, H.A., Lynd, F.T., Masoro, E.J. & Yu, B.P. (1980). Changes in adipose mass and cellularity through the adult life of rats fed *ad libitum* or a life-prolonging restricted diet. Journal of Gerontology, **35**, 827-835.

Butcher, R.L. & Page, R.D. (1981). Role of the aging ovary in cessation of reproduction. In: M.B. Schwartz & M. Hunzicher-Dunn (eds.), Dynamics of Ovarian Function, pp. 253-271. New York: Raven Press.

Butcher, R.L. & Pope, R.S. (1979). Role of estrogen during prolonged estrous cycles of the rat on subsequent embryonic death or development. Biology of Reproduction, **21**, 491-495.

Chvapil, M. & Hrůza, Q. (1959). The influence of aging and undernutrition on chemical contractility and relaxation of collagen fibres in rats. Gerontologia, **3**, 241-252.

Comfort, A. (1979). The Biology of Senescence, 3rd edn. London: Churchill Livingstone.

Cooper, B., Weinblatt, F. & Gregerman, R.I. (1977). Enhanced activity of hormone-sensitive adenylate cyclase during dietary restriction in the rat. Dependence on age and relation to cell size. Journal of Clinical Investigation, **59**, 467-474.

Davies, T., Bales, C.W. & Beauchene, R.E. (1983). Differential effects of dietary calorie and protein restriction in the aging rat. Experimental Gerontology, **18**, 427-435.

Deyl, Q., Juricova, M., Rosmus, J. & Adam, M. (1971). The effect of food deprivation on collagen accumulation. Experimental Gerontology, **6**, 383-390.

Enesco, H.E. & Kruk, P. (1981). Dietary restriction reduces fluorescent age pigment accumulation in mice. Experimental Gerontology, **16**, 357-361.

Enesco, H.E. & Samborsky, J. (1983). Liver polyploidy: influence of age and of dietary restriction. Experimental Gerontology, **18**, 79-87.

Everitt, A.V. (1982). Nutrition and the hypothalamic-pituitary influence on aging. In: G.B. Moment (ed.), Nutritional Approaches to Aging Research, pp. 245-256. Florida: CRC Press, Inc.

Everitt, A.V., Porter, B.D. & Steele, M. (1981). Dietary, caging and temperature factors in the ageing of collagen fibres in rat tail tendon. Gerontology, **27**, 37-41.

Frolkis, V.V. (1982). Aging and Life-Prolonging Processes. Vienna: Springer-Verlag.

Frolkis, V.V., Bogatskaya, L.N., Stupina, A.S. & Verzhikovskaya, N.V. (1976). Comparative characterization of some influences of life prolongation. In: Biological Abilities of Increasing the Animal's Life Span, pp. 138-150. Kiev: Institute of Gerontology.

Fugo, N.W. & Butcher, R.L. (1966). Overripeness and the mammalian ova. I. Overripeness and early embryonic development. Fertility and Sterility, **17**, 804-813.

Fugo, N.W. & Butcher, R.L. (1971). Effects of prolonged estrous cycles on reproduction in aged rats. Fertility and Sterility, **22**, 98-101.

Gahan, P.B., Merry, B.J. & Middleton, J. (1985). Dietary restriction and polyploidization in rat hepatocytes. Submitted.

Giles, J.S. & Everitt, A.V. (1967). The role of the thyroid and of food intake in the aging of collagen fibres. Gerontologia, **13**, 65-74.

Herlihy, J.T. & Yu, B.P. (1980). Dietary manipulation of age-related decline in vascular smooth muscle function. American Journal of Physiology, **238**, H652-H653.

Holehan, A.M. (1984). The effect of ageing and dietary restriction upon reproduction in the female CFY Sprague Dawley rat. Ph.D. Thesis, University of Hull.

Hollander, C.F. (1979). Pathophysiology of ageing in animal models. In: J. Crooks & I.H. Stevenson (eds.), Drugs and the Elderly, pp. 15-22. London: MacMillan Press Ltd.

Hrůza, Q. & Hlavackova, V. (1969). Effect of environment temperature and undernutrition on collagen aging. Experimental Gerontology, **4**, 169-175.

Huang, H.H., Steger, R.W., Bruni, J.F. & Meites, J. (1978). Patterns of sex steroid and gonadotropin secretion in aging female rats. Endocrinology, **103**, 1853-1859.

Kawada, T., Fujisawa, T., Imai, K. & Ogata, K. (1977). Effects of protein deficiency on the biosynthesis and degradation of ribosomal RNA in rat liver. Journal of Biochemistry, **81**, 143-152.

Leto, S., Kokkonen, G.C. & Barrows, C.H. (1976). Dietary protein, lifespan and biochemical variables in female mice. Journal of Gerontology, **31**, 144-148.

Levin, P., Janda, J.K., Joseph, J.A., Ingram, D.K. & Roth, G.S. (1981). Dietary restriction retards the age-associated loss of rat striatal dopaminergic receptors. Science, **214**, 561-562.

Liepa, G.U., Masoro, E.J., Bertrand, H.A. & Yu, B.P. (1980). Food restriction as a modulator of age-related changes in serum lipids. American Journal of Physiology, **238**, E253-E257.

Lloyd, T. (1984). Food restriction increases lifespan of hypertensive animals. Life Sciences, **34**, 401-407.

Lloyd, T. & Boyd, B. (1981). Development and regulation of hypertension in the spontaneously hypertensive rat: enzymatic and nutritional studies. In: E. Usdin, N.Weiner & M.B.H. Youdim (eds.), Function and regulation of monoamine enzyme: Basic and clinical aspects, pp. 843-854. London: Macmillan.

Lu, J.K.H. (1983). Changes in ovarian function and gonadotropin and prolactin secretion in aging female rats. In: J. Meites (ed.), Neuroendocrinology of Aging, pp. 103-122. New York: Plenum Press.

Masoro, E.J., Bertrand, H., Liepa, G. & Yu, B.P. (1979). Analysis and exploration of age-related changes in mammalian structure and function. Federation Proceedings. Federation of American Societies for Experimental Biology, **38**, 1956-1961.

Masoro, E.J., Yu, B.P., Bertrand, H.A. & Lynd, F.T. (1980). Nutritional probe of the aging process. Federation Proceedings. Federation of American Societies for Experimental Biology, **39**, 3178-3182.

Masoro, E.J., Yu, B.P. & Bertrand, H.A. (1982). Action of food restriction in delaying the aging process. Proceedings of the National Academy of Sciences, USA, **79**, 4239-4241.

McCay, C.M., Crowell, M.F. & Maynard, L.A. (1935). The effect of retarded growth upon the length of lifespan and upon the ultimate body size. Journal of Nutrition, **10**, 63-69.

McCay, C.M., Sperling, G. & Barnes, L.L. (1943). Growth, aging, chronic diseases and lifespan in rats. Archives of Biochemistry, **2**, 469-479.

Merry, B.J. & Holehan, A.M. (1979). Onset of puberty and duration of fertility in rats fed a restricted diet. Journal of Reproduction and Fertility, **57**, 253-259.

Merry, B.J. & Holehan, A.M. (1981). Serum profiles of LH, FSH, testosterone and 5α-DHT from 21 to 1000 days of age in *ad libitum* fed and dietary restricted rats. Experimental Gerontology, **16**, 431-444.

Merry, B.J. & Holehan, A.M. (1985a). *In vivo* DNA synthesis in the dietary restricted long-lived rat. Experimental Gerontology, **20**, 15-28.

Merry, B.J. & Holehan, A.M. (1985b). The effect of refeeding on subsequent survival and DNA synthesis in the dietary restricted long-lived rat. In preparation.

Merry, B.J. & Phillips, J.G. (1981). Basic Gerontology. In: M. Green (ed.), Clinics in Endocrinology and Metabolism, Endocrinology and Ageing, pp. 3-22. Eastbourne: W.B. Saunders.

Merry, B.J., Holehan, A.M. & Philips, J.G. (1985). Modification of reproductive decline and lifespan by dietary manipulation in female CFY Sprague-Dawley rats. 9th International Symposium of Comparative Endocrinology, Hong Kong, vol. 1, pp. 621-624. Hong Kong: University Press.

Mikamo, K. (1970). Anatomic and chromosomal anomalies in spontaneous abortion. American Journal of Obstetrics and Gynecology, **106**, 243-254.

Miller, D.S. & Payne, P.R. (1968). Longevity and protein intake. Experimental Gerontology, **3**, 231-234.

Miller, A.E. & Riegle, G.D. (1979). Endocrine factors associated with the initiation of constant estrous in aging female rats. Federation Proceedings. Federation of American Societies for Experimental Biology, **38**, 1248.

Moment, G.B. (ed.) (1982). Nutritional Approaches to Aging Research. Florida: CRC Press Inc.

Nakagawa, I., Sasaki, A., Kajimoto, T., Fukuyama, T., Suzuki, T. & Yamada, E. (1974). Effect of protein nutrition on growth, longevity and incidence of lesions in the rat. Journal of Nutrition, **104**, 1576-1583.

Nandy, K. (1982). Neuroimmunology and the aging brain. In: S. Hoyer (ed.), The Aging Brain: Physiological and Pathophysiological Aspects. Experimental Brain Research, Suppl. 5, pp. 121-126. Berlin: Springer-Verlag.

Nolen, G.A. (1972). Effects of various restricted dietary regimes on the growth, health and longevity of albino rats. Journal of Nutrition, **102**, 1477-1494.

Osborne, T.B., Mendel, L.B. & Ferry, E.L. (1917). The effect of retardation of growth upon the breeding period and duration of life in rats. Science, **45**, 294-295.

Payne, P. (1979). Ageing and Nutrition. In: J. Crooks & I.H. Stevenson (eds.), Drugs and the Elderly, pp. 39-47. London: Macmillan.

Peluso, J.J., England-Charlesworth, C. & Hutz, R. (1980). Effect of age and of follicular aging on the preovulatory oocyte. Biology of Reproduction, **22**, 999-1005.

Ross, M.H. (1959). Protein, calories and life expectancy. Federation Proceedings. Federation of American Societies for Experimental Biology, **18**, 1190-1207.

Ross, M.H. (1964). Nutrition; disease and length of life. In: G.E.W. Wolstenholme & M. O'Connor (eds.), Diet and Bodily Constitution. Ciba Foundation Study Group, No. 17, pp. 91-103. Boston: Little Brown.

Ross, M.H. (1966). Life expectancy modification by change in dietary regimen of the mature rat. Proceedings of the 7th International Congress of Nutrition, **5**, 35-38. New York: Pergamon Press.

Ross, M.H. (1969). Aging, nutrition and hepatic enzyme activity patterns in the rat. Journal of Nutrition, Suppl. 1, **97**, 563-602.

Ross, M.H. (1972). Length of life and caloric intake. American Journal of Clinical Nutrition, **25**, 834-838.

Ross, M.H. & Bras, G. (1965). Tumor incidence patterns and nutrition in the rat. Journal of Nutrition, **87**, 245-260.

Ross, M.H. & Bras, G. (1971). Lasting influence of early caloric restriction on prevalence of neoplasms in the rat. Journal of the National Cancer Institute, **47**, 1095-1113.

Ross, M.H. & Bras, G. (1973). Influence of protein under and over nutrition on spontaneous tumor prevalence in the rat. Journal of Nutrition, **103**, 944-963.

Ross, M.H. & Bras, G. (1974). Dietary preference and diseases of age. Nature, **250**, 263-265.

Ross, M.H. & Bras, G. (1975). Food preference and length of life. Scence, **190**, 165-167.

Ross, M.H., Bras, G. & Ragbeer, M.S. (1970). Influence of protein and caloric intake upon spontaneous tumor incidence of the anterior pituitary gland of the rat. Journal of Nutrition, **100**, 177-189.

Sacher, G.A. (1977). Life table modification and life prolongation. In: C.E. Finch & L. Hayflick (eds.), Handbook of the Biology of Aging, pp. 582-638. New York: Van Nostrand Reenhold Company.

Samuels, L.T. (1946). The relation of the anterior pituitary hormones to nutrition. Recent Progress in Hormone Research, **1**, 147.

Srivastava, U., Ganguli, P.K., Brasseur, R. & Gyenes, L. (1978). The metabolism of liver RNA in adult rats subjected to prolonged dietary restriction during the period of growth and development. Nutritional Reports International, **17**, 367-375.

Stuchlikova, E., Juricova-Horakova, M. & Deyl, Z. (1975). New aspects of the dietary effect of life prolongation in rodents. What is the role of obesity in aging? Experimental Gerontology, **10**, 141-144.

Swann, P.F. & Magee, P.N. (1968). Nitrosamine induced carcinogenesis. Biochemical Journal, **100**, 39-47.

Tannenbaum, A. & Silverstone, H. (1949). The influence of the degree of caloric restriction on the formation of skin tumours and hepatomas in mice. Cancer Research, **1**, 452-501.

Tannenbaum, A. & Silverstone, H. (1953). Nutrition in relation to cancer. Advances in Cancer Research, **1**, 452-501.

Tucker, S.M., Mason, R.L. & Beauchene, R.E. (1976). Influence of diet and feed restriction on kidney function of aging male rats. Journal of Gerontology, **31**, 264-270.

Visscher, M.B., King, J.T. & Lee, Y.C.P. (1952). Further studies on influence of age and diet upon reproductive senescence in strain A female mice. American Journal of Physiology, **170**, 72-76.

Walford, R.L., Liu, R.K., Gerbase-Delima, M., Mathies, M. & Smith, G.S. (1974). Longterm dietary restriction and immune function in mice. Mechanisms of Ageing and Development, **2**, 447-454.

Weindruch, R.H. & Walford, R.L. (1982). Dietary restriction in mice beginning at 1 year of age: effect on life-span and spontaneous cancer incidence. Science, **215**, 1415-1418.

Weindruch, R.H., Kristie, J.A., Cheney, K.E. & Walford, R.L. (1979). Influence of controlled dietary restriction on immunologic function and aging. Federation Proceedings. Federation of American Societies for Experimental Biology, **38**, 2007-2016.

Woodhead, A.D., Merry, B.J., Cao, E-H., Holehan, A.M. & Carlson, C. (1984). The levels of O^6-Methylguanine-acceptor protein in tissues of rats throughout their lifespan. Submitted.

Yu, B.P., Bertrand, H.A. & Masoro, E.J. (1980). Nutrition-aging influence of catecholamine-promoted lipolysis. Metabolism Clinical and Experimental, **29**, 438-444.

Yu, B.P., Masoro, E.J., Murata, I., Bertrand, H.A. & Lynd, F.T. (1982). Lifespan study of SPF Fischer 344 male rats fed *ad libitum* or restricted diets: Longevity, growth, lean body mass and disease. Journal of Gerontology, **37**, 130-141.

CUSTOMARY PHYSICAL ACTIVITY IN THE ELDERLY

J. M. PATRICK

Department of Physiology & Pharmacology, University Medical School
Queen's Medical Centre, Clifton Boulevard, Nottingham, U.K.

INTRODUCTION

The companion papers presented in this volume give full accounts of the sheer size of the elderly population in various societies, the state of the physical and mental health of elderly people, and their social and nutritional requirements. Concern has been expressed about both the *quality* of life and the *length* of life, and those fundamentally different ideas are also reflected within the field of activity measurement by a parallel interest in both the *intensity* and the *duration* of physical work. In younger subjects we are usually interested in measuring customary physical activity (or inactivity) because of its supposed relation with risk of cardiovascular disease, with energy balance and obesity, and with physiological capacities (particularly capabilities for sport and work). In older people, our attention focusses very sharply on the potentially vicious spiral of inactivity leading to physical deconditioning and thence, via loss of physiological capacity, to a further decline in activity (Bassey, 1978; Bassey et al, 1983; Evans, this volume). Ultimately the penalty is dependence and institutionalization; and the relevant criterion for physiologists, medical practitioners, epidemiologists and health care planners alike is not simply life expectancy but "active life expectancy" (Katz et al, 1983) with its connotations of independence and well-being. These are exemplified by the reputed life-styles of the centenarians of the Caucasus (Medvedev, this volume).

We rather expect elderly people who have lived and worked in an industrial society to become less active in retirement, irrespective of their actual capacities and capabilities. The phrases "put your feet up" and "now you can take it easy" come freely when we talk to people of our parents' generation. Such cultural influences on our activity patterns are very strong, and these attitudes will take some overturning by the new patterns arising from our recent awareness of the benefits of aerobic exercise in its various forms. While the dangers of unaccustomed excess must not be neglected, it is still true that the

opportunities for sensible, appropriate and beneficial physical activity are available to all. How are these opportunities used by the elderly? Can we measure their customary physical activity? Can we show how much less than it is in younger subjects, and whether or not it is sharply reduced at retirement? This whole subject was extensively reviewed by Shephard in 1978, but it is clear that there was rather little hard data to rely on then. Reports of conferences organised by the American College of Sports Medicine and the Institute de la Vie contain accounts of the role of exercise in ageing (Sidney, 1981; Hodgson & Buskirk, 1981), and there have been a few other reports scattered through the literature. Many authors have stressed the difficulty of such measurements: if they are not done well, the conclusions cannot be relied on (LaPorte et al, 1982). This paper will first review the methods and approaches available, and then discuss the main findings of key studies from Britain and North America.

METHODS OF ESTIMATING CUSTOMARY PHYSICAL ACTIVITY IN THE ELDERLY

The four main methods applicable in any age-group are listed in Table 1, and examples of their use in the elderly are given below. Broadly speaking, the more objective we require the data to be, the more expensive they are to collect and analyse (Bassey & Fentem, 1980). All methods raise the same serious

Table 1. The Main Methods of measuring Customary Physical Activity

A. Retrospective	1.	Questionnaire: interview
B. Prospective	2.	Diary annotation by (a) subject (b) observer supplemented by direct measurement in the laboratory of the strain of a sample of activities
	3.	Body-borne instruments (ECG; $\dot{V}O_2$; movement of trunk or limbs) with (a) body-borne recorders or telemetry (b) on-line data-reduction (c) off-line analysis
C. Indirect	4.	Energy intake measurement using methods 1 or 2, assuming subjects are in energy balance.

questions about reliability and validity, especially as regards sampling the subjects and the periods of observation, and relating the particular activities that are recorded to those that are relevant to the objectives of the study. None are entirely satisfactory.

The questionnaire or interview method (A.1 in Table 1) asks questions in a more or less structured way about a subject's recent physical activities. The apparent convenience and simplicity account for its popularity and enable large numbers to be studied, but elderly subjects will have particular difficulties in remembering, in hearing and understanding the interviewer, or in filling in forms; and they will probably be more liable to make too much or too little effort to provide the answers expected. It has often been demonstrated that retrospective questionnaire responses exaggerate the intensity and duration of the physical activity undergone. It is important to validate any such method against a more objective one.

Prospective methods (Table 1, B) are generally preferable, though they are normally too expensive for the study of large subject groups, and they introduce the real possibility that the very pattern of activity they purport to measure is being altered by the act of measuring. The first of these methods is the collection of minute-by-minute descriptions of current activities using diary cards which are filled in frequently either by observers or by the subjects themselves. Failing comprehension or manual dexterity limit the use of the latter method in the elderly, but direct observation is no more difficult in this age-group (it is never easy) than for younger subjects. Simultaneous observation of a group of old people restricted to a communal home would be relatively economical. This approach provides an accurate measure of the duration of various activities, but their intensity can only be gauged by supplementary experimental determinations of the increases in either oxygen uptake or heart-rate associated with the particular activity in question. These are open to the objections that also apply to the third method, *viz* continuous recording of physiological indices of the strain of physical activity.

Miniaturization of laboratory analysers has allowed the development of body-borne devices that can provide at least semi-continuous measurement of heart-rate (from ECG electrodes), oxygen uptake (using the Oxylog which measures the oxygen content and throughput of expired air), ventilation (using thoracic transducers) and movement of the trunk or limbs (using accelerometers or other motion sensors). The data can be telemetered to a fixed laboratory (Patrick, 1983) or stored on a body-borne recorder. The major advantages of

this approach are that (i) the equipment (apart from the Oxylog) is unobtrusive enough to allow the subject to forget that he is being monitored; (ii) information is being collected about the subject's natural activities rather than about a formal simulation in the laboratory; and (iii) the subject is free from the restraints and tethering usually associated with laboratory measurement. These recent advances heralded a new era in this field, but have presented experimenters with the problem of data-reduction. The handling of more than 10^5 heart beats recorded per subject-day on a cassette-tape requires an enormous investment in computer hardware if their sequence is to be preserved in order to provide an analysis of the pattern of heart-rate during the day. The E-cell accumulator of the SAMI which counts only the daily total of heart beats is simple enough, but it cannot distinguish intensity from duration of activity, nor can the inevitable artefacts be identified and edited out from the record. Elderly people can wear tape-recorders, pedometers or accelerometers without difficulty if care is taken to affix and secure them properly, and due attention is given to fragile and inelastic skin. The heart-rate record needs to be interpreted, as with younger subjects, in terms of the response to some standard calibrating activity (see below) in order to allow for differences between subjects' physical condition and for the attenuation of autonomic responses to physical activity.

Finally, if body weight and its components are stable and the subjects are presumed to be in energy balance, then measurements of energy intake could be used to estimate energy expenditure. The methods available for this are open to exactly the same objections as apply to the direct approach, and seem to offer no particular advantage unless nutrition itself is being studied.

So the experimenter wishing to measure customary physical activity in the elderly has available a varied battery of methods of differing precision, objectivity and convenience. He should choose carefully. His choice might well determine how successful he will be in coping with the broader problems of variation between and within subjects. The design of the study depends on the precision and convenience of the method of measurement which tend to be inversely related. Longitudinal and intervention studies, which reduce the variance considerably, are to be preferred, but they require long-term funding and a stable group of patient investigators who are highly systematic and impeccably tidy in their record-keeping.

VARIABILITY BETWEEN AND WITHIN SUBJECTS

Even among subjects over 65 years old it is clear that there is a wide range of levels of activity; and cultural, seasonal, and medical factors have a profound influence. The study-group needs therefore to be carefully described and of appropriate size. Bias is easily introduced when volunteer subjects are recruited: the more active, the more outgoing and the more hypochondriacal tend to select themselves. If subjects are excluded on medical grounds either before or after volunteering, the relation between the study-group and the population as a whole can be distorted (see Borkan, this volume; and Bailey, this volume). Dropout in longitudinal studies, which is high anyway, tends to leave the healthier subjects to be included in the final analysis. In any intervention study, the volunteer subjects must be allocated randomly to control and test groups to allow for biases such as these.

The day is the standard unit of time for studies of activity, but this may tempt us to ignore the day-to-day variation. Stunkard (1960) provocatively likened the day-to-day pattern of activity over 23 days in young adults to a sequence of random numbers, and then pointed out the weekly rhythm embedded in the former. Our own data in Figure 1, also obtained with pedometers, suggests that weekend days are neither more nor less active than weekdays

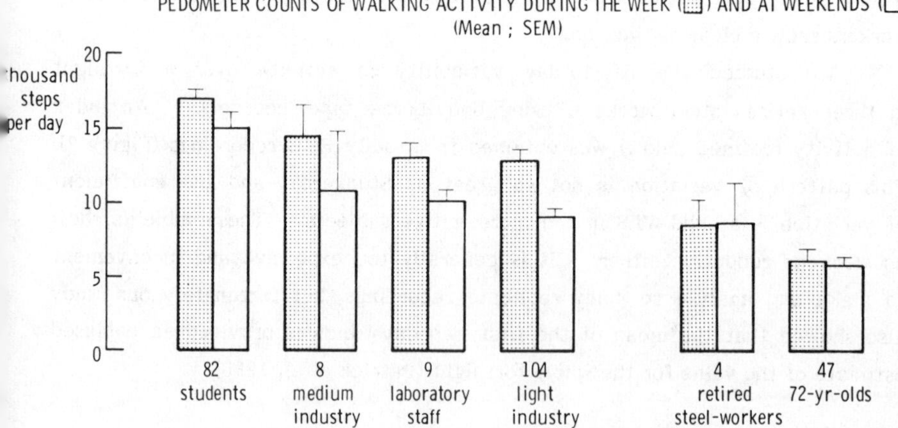

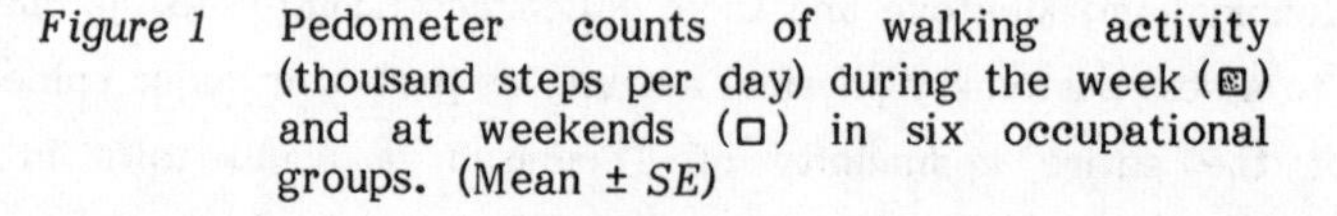

Figure 1 Pedometer counts of walking activity (thousand steps per day) during the week (▒) and at weekends (□) in six occupational groups. (Mean ± *SE*)

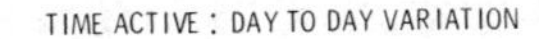

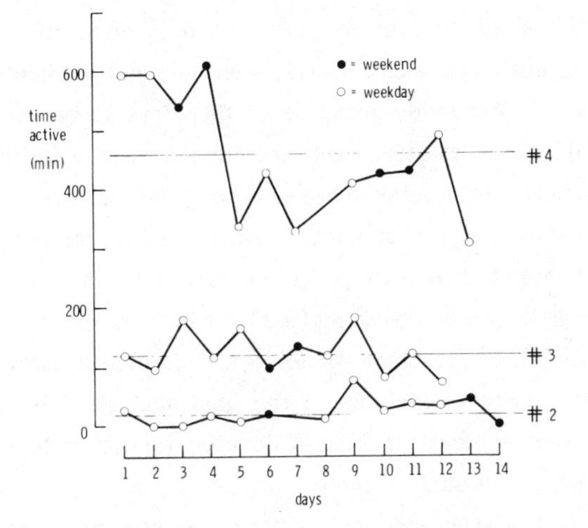

Figure 2. Fourteen consecutive days' activity in three retired men, and their average values for the fortnight. The activity index is the number of minutes spent at or above an individual reference heart-rate corresponding to walking at 4.8 $km.hr^{-1}$ (less 5 $bt.min^{-1}$) for periods of at least 7.5 min at a time (see also Figure 6).

for retired steel workers and for a sample of elderly people aged 72 years from a single practitioner's list, while students, laboratory workers and factory workers show a clear distinction.

We studied the day-to-day variability in activity over a fortnight in three retired steel workers, using body-borne tape-recorders. An index of activity (defined below) was obtained from daily ECG recordings (Figure 2). This pattern of variation is not so great as Stunkard's, and the coefficient of variation is around 50% in the more active subjects. These subjects show no weekend reduction either. It is generally too expensive and inconvenient to make and analyse so many replicate recordings, but fortunately our study also showed that the mean of the first two days' indices provided an unbiased estimate of the value for the whole fortnight (Patrick et al, 1985).

PREVIOUS STUDIES

Cunningham, Montoye and their colleagues (1968) used the questionnaire method to assess customary physical activity as part of a major epidemiological study of the entire community of Tecumseh, a small town in Michigan.

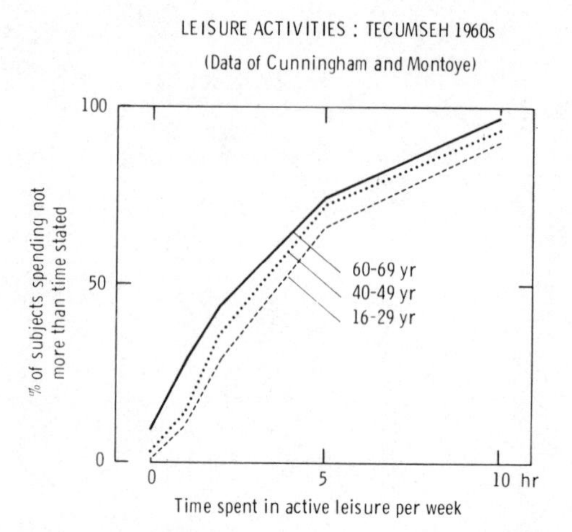

Figure 3. Cumulative frequency distribution showing the percentage of men in each of three age-groups who spent not more than the stated number of hours per week in all active leisure pursuits. (Data of Cunningham et al, 1968.)

The data were collected in 1965 on some 1,650 adult men using an 'Activity Recall Record' which combined a self-administered questionnaire and a personal interview. No objective measurements were made. They first established the number of hours per week (averaged over the whole year) spent in leisure activities. For the sake of clarity, Figure 3 only includes the youngest adult group (16-29 yr; n = 360), the middle-aged (40-49 yr; n = 431) and the oldest group reported (60-69 yr; n = 149). The other groups lie in between. This cumulative distribution curve, giving the percentage of each age-group active for not more than the number of hours stated, shows first that on average individuals spent about 3 hr per week in leisure activities, while the most active 20% spent an hour per day. The percentage taking *no* active leisure increases a little with age, and the activity undertaken by the median individuals fell from 3.5 hr per week (30 min per day), through 3 hr at 45 yr to 2.5 hr per week in the seventh decade. As the analysis excluded occupational activities this steady trend suggests a substantial ageing decline. However, if we look specifically at activities in which subjects can engage at any age, then the picture looks less gloomy. Figure 4 shows that more rather than fewer of the older subjects spent time in walking and in gardening. The

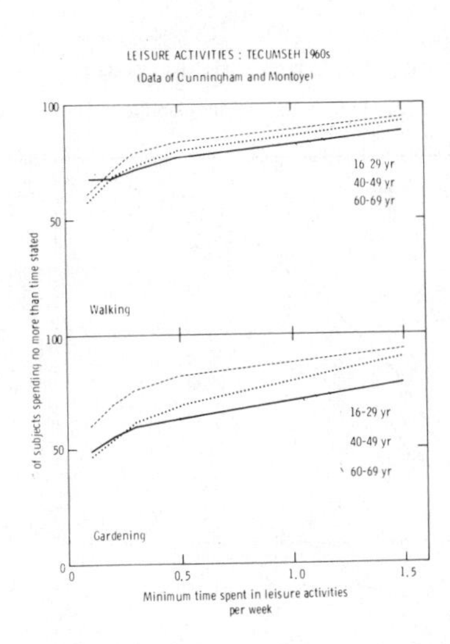

Figure 4. Cumulative frequency distribution showing the percentage of men in each of three age-groups who spent not more than the stated number of hours per week in walking and gardening activities. (Data of Cunningham et al, 1968.)

differences are measured in minutes per week rather than in hours, but this finding does at least remind us that there are types of leisure activity in which it is common and natural for older people to spend their time. Other popular activities were swimming, hunting, dancing and fishing in descending order of intensity. It is notable that in 1965 jogging featured nowhere on the list, and cycling had very few adherents. Paffenbarger et al (1978) have also used the questionnaire method in a similar study.

The first systematic study of the activity of elderly people was pioneered by John Durnin in Glasgow in the 1950s (Durnin et al, 1957, 1961a, 1961b). He used the second method (see Table 1), getting the subjects to complete a minute-by-minute diary card for a whole week. That takes about seven thousand entries but it does record the duration and type of every activity undertaken. The method, which was used as part of a study of energy expenditures, had been thoroughly validated in younger subjects. The data are summarized in Table 2, which shows the time spent per day in various active pursuits, i.e. excluding bed-rest and sitting which together account for about two-thirds of the day. Four groups of women were studied, two of them being matched pairs of mothers and daughters. The latter were employed serving

Table 2. Number of minutes spent per day by four groups of women in various activities (Data of JVGA Durnin and colleagues, 1957-61)

	Young adult shop workers n=12	Middle-aged housewives n=12	Elderly with families n=21	Elderly living alone n=15
Age range (yrs):	16-25	45-61	54-66	60-69
Personal necessities	36	24	55	29
"Pottering"	375	420	379	305
Walking outdoors including shopping	84	63	92	89
Total time on feet	501 (8.5 hr)	507 (8.5 hr)	526 (8.5 hr)	423 (7.0 hr)

customers in a department store while their mothers, like the women in the other two groups, were housewives. These data show that although women were on their feet for 7 to 8 hours, only 60 to 90 minutes were spent in walking of any material duration. In further analyses Durnin et al showed that in these women only one hour's activity per day could be classed as moderate or harder (and 59 of those 60 minutes were *not* harder). The second important finding is that the middle-aged and elderly women with families were as active as their daughters: no inevitable decline with age here. Although the elderly women living alone did seem to be less active in the home, they spent the same time in walking and shopping activities which are probably the most important from the point of view of maintaining physical condition. A similar approach was used by Sidney & Shephard (1977) in a small study of elderly Canadians entering a pre-retirement fitness training programme.

These studies have relied on interview and observational methods; the rest use more objective measurements. Perhaps the simplest body-borne device that records some function of activity is the motion sensor. The pedometer is just one of a wide range of such devices, and we have used it extensively in Nottingham (Irving & Patrick, 1982). Essentially it counts those vertical 'jogs' whose peak acceleration exceeds the instrument's threshold. When pedometers that respond over a low range of peak accelerations are selected from the batch and calibrated with a sinusoidal movement representing the acceleration occurring at the hip during normal walking, they provide

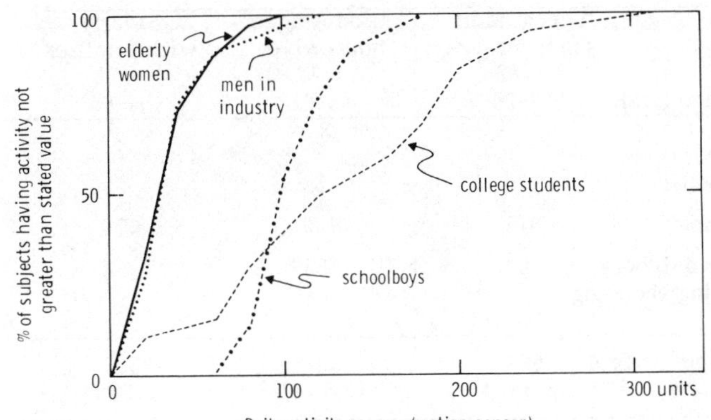

Figure 5. Cumulative frequency distribution showing the percentage of subjects in each of four occupational groups whose daily activity, measured by a motion sensor, was no more than the stated value. (Data of LaPorte et al, 1982.)

a reasonable estimate of the number of steps taken. Figure 2 showed that pedometers can discriminate activity levels between occupational and retired groups, and between weekday and weekend levels. We have also demonstrated that they can detect the extra 20 minutes of walking undertaken by a large group of 55 to 60-year-old subjects during an intervention study (Bassey et al, 1983).

Laporte and his colleagues in Pittsburgh have used a somewhat more sophisticated device called a Large Scale Integrator (LSI) which is essentially a mercury switch responding to slight perturbations in the position of the trunk. Like the pedometer, it is mounted on the hip, and a count for each of several days gives an index of customary physical activity (Laporte et al, 1982). Figure 5 summarizes their data in four subject groups, and shows a clear difference in activity levels between schoolboys and college students on the one hand, and elderly women and middle-aged men on the other. The men, steel-mill labourers in their last years of work, are very similar to the ones we have studied in Nottingham (see below); and it is striking that their activity is *not* different from the elderly ladies. The authors point out that despite the fact that these men are judged to be sedentary

on the basis of this evidence, two-thirds of their total activity is performed at work. This study again shows how difficult it is to demonstrate a clear decline in activity with age in later years. A further illustration of this is provided by another study that used a similar device called a 'Patient Activity Monitor' attached to the wrist of only 14 subjects (Renfrew et al, 1984).

THE NOTTINGHAM STUDY

A team in Nottingham has been studying the physical activity of elderly people over the last 12 years. We have been particularly interested in the relation between customary physical activity and both body composition (Patrick et al, 1982) and physical condition (Bassey et al, 1982; Patrick et al, 1983), and we centred our study on the time of retirement from work. We believed that this was the nearest we could get to an experimental intervention, as it seemed likely that the decisive change in life-style at retirement would be associated with a reduction in physical activity. In parallel, we have studied the effects of an actual intervention in the opposite direction, a 12-week walking programme, in volunteer workers in their last five years before retirement (Bassey et al, 1983).

We used the body-borne tape-recorder as our main tool, storing the continuous ECG waveform together with a footfall signal for two separate full days in the same week (Bassey et al, 1980). This entails the application of electrodes to the skin of the chest and a carbon pad to the heel, and the leads are brought out through the clothing to a miniature 0.4 kg cassette tape-recorder (Oxford Instruments) worn unobtrusively on the belt. The only activities which cannot be undertaken while wearing the recorder are showering and bathing: most subjects put the recorder, still connected, under the pillow at night. We had no complaints that the measuring procedure restricted the subjects in any way, and our results suggest that the two days of recording were not systematically different from 12 subsequent days: the subjects were not putting off activities to a later date, or trying unduly to impress us.

Three groups have been studied before retirement from work and then again one year later. There were about twenty subjects in each group, the first of which were male manual workers in a heavy industry: a steel and concrete plant near Nottingham. The study was undertaken before the recession brought redundancies and early retirement: these men worked until their 65th birthday though many had been moved to lighter work during

their 60s. The second and third groups were men and women working on the shop floor in a light industrial company in Nottingham. They retired in their early 60s and had not changed their occupation in their last years of work. Recruitment was done on a voluntary basis through the personnel and pensions offices at the factories, and it was unlikely that the sample was any more or less representative than in other studies of this nature. The first measurements were made during the last month at work, the tape-recorders being fitted on two separate days at the start of the shift and collected 24 hours later each time. The second pair of measurements were made a year later, and we visited the subjects in their homes.

Analysis of the data on the cassette-tapes has proved to be a substantial task, and we have already gone through two generations of play-back and A-D converting systems, and two main-frame computers. Other systems provide either mean heart-rates over a day, a shift, or an hour; or the time spent with heart-rates in a handful of ranges, e.g. below 60, 60-79, 80-99, 100-119, and over 120 bt. min^{-1} (e.g. Ilmarinen & Rutenfranz, 1979). Our analysis, in contrast, had been predicated on three main premises. The first is that the physical activity relevant to maintenance or improvement in physical condition needs to be of some minimum duration (say 2 min) and some minimum intensity to be at all effective (deVries, 1971), so the recording and analysis must preserve information about the duration of each bout of heart-rate elevation. The second premise is that such periods do not occur very often or persist for very long, so they must not be averaged away by pooling with inactive periods. The third is that the actual heart-rate elevation induced by such activity will itself depend on the physical condition of the subject who therefore needs to be individually calibrated.

On each occasion, therefore, we 'calibrate' the subject by getting him to perform a standard self-paced walking-test along level corridors at three different speeds (Bassey et al, 1982). This provides, by interpolation, the heart-rate required by that subject for a standard walking-speed of 4.8 $km.hr^{-1}$ (i.e. 3 $miles.hr^{-1}$). The procedure can also be regarded as a test of physical condition or fitness. From this heart-rate we subtract 5 $bt.min^{-1}$ to allow for test-retest variation: this is then the reference heart-rate. We are only concerned with elevations above this level, representing an intensity of activity at least as great as walking at a moderate pace. Our computer analysis (Figure 6) accumulates all periods of heart-rate elevation lasting at least 7.5 minutes ('sustained activity'), and calculates the average

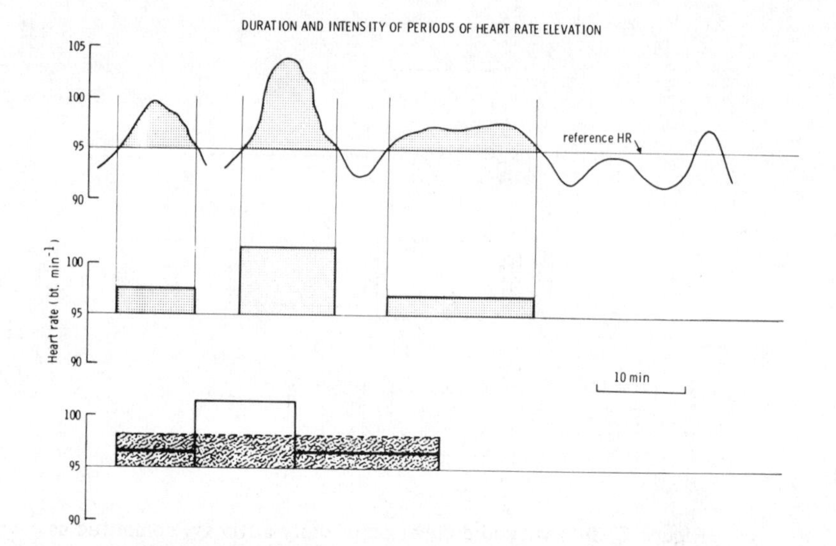

Figure 6. Diagrammatic representation of the derivation by computer of an index of customary physical activity from a continuous record of heart-rate during a day. The upper panel shows three periods of heart-rate elevation above the reference level long enough to qualify in this analysis. These periods are stored as three blocks (middle panel) preserving their duration and area (duration x intensity). The whole day's index is the shaded block in the lower panel which has the same total duration and same total area as all the blocks in the middle panel.

intensity as well as the intensity x duration product which turns out to be a count of heart beats in excess of those that would have occurred had the subject spent the whole day at a lower intensity (Bassey et al, 1983). The intensity and duration of a whole day's activity can be represented as a single block showing the total duration of all the periods along the time axis, and the average heart-rate elevation over the reference level on the ordinate. The area of the block represents the intensity x duration product for the day. Every subject has *some* activity at the level of sustained walking while wearing the recorder for two days, although Figure 2 shows that the least active subjects have some non-walking days in a fortnight. This analysis does not take account of the timing of activity within the day, but there is no evidence as yet that this is important.

The results are summarized in Figure 7 which gives composite data for the three groups of subjects both before and after retirement from work.

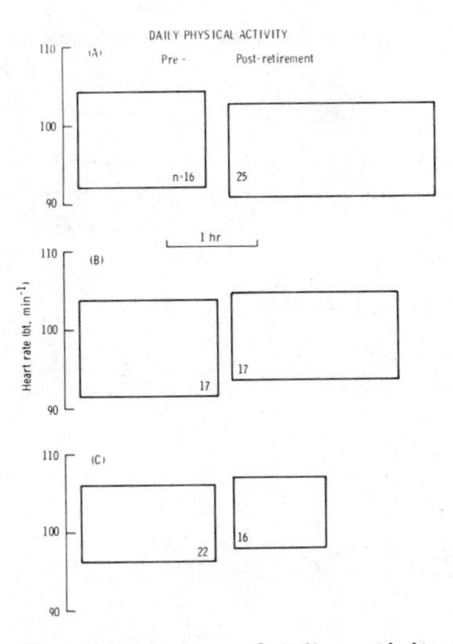

Figure 7. Intensity and duration of daily activity, computed as shown in Figure 6, for three groups of subjects before and after retirement from work. Group A, men in a medium-heavy industry; Groups B and C, men and women in a light industry. The baseline of each block is at the group's mean reference heart-rate.

The horizontal scale is condensed by comparison with Figure 6. Each block represents about five periods of sustained activity, so on average the periods (adding up to an hour or more) were considerably longer than the criterion value of 7.5 minutes.

For the men retiring from the medium engineering factory (A) and from the light engineering factory (B), there are no material or significant differences for any of the variables represented: i.e. duration, intensity, duration x intensity product or number of periods were not changed after retirement. However, for the women retiring at about the age of 60 yr (Group C), there does seem to be a noticeable reduction in daily physical activity amounting to about 40% of the pre-retirement values when considering these variables and others analysed but not shown here.

CONCLUSIONS

We have seen that the measurement of customary physical activity is difficult and complicated at the best of times, and that there are some extra problems with some methods in the elderly. The variability between and within subjects is large, and therefore great care is needed with sampling

to ensure adequate representation of the whole range of subjects and the whole range of activities. Precision will be low, and longitudinal paired measurements are likely to be the most efficient at detecting differences due to age, change in occupation, taking up exercise programmes, etc.

The broad conclusion emerging from the questionnaire or diary methods is that a modest reduction in total time spent in active pursuits *is* apparent when comparing older with younger subjects, but that the elderly still spend time in simple beneficial activities like walking and gardening. This suggests that exercise programmes based on these activities, rather than on sport, jogging or calisthenics, are likely to be successful in persuading older people to maintain their physical activity and thus their physical condition.

The objective methods, using body-borne motion sensors and tape-recorders are beginning to provide hard evidence about intensity of activity as well as duration. Rather little change of activity is seen at retirement in men, even those leaving a "heavy" industry. This suggests that leisure pursuits can provide as much physical activity as was required in the last years at work, though this was not necessarily very much. The reduction in activity in women on retirement warns us of a group who may be in particular need of health education.

ACKNOWLEDGEMENTS

I am glad to have had the opportunity of working over several years with Professor P. H. Fentem and Dr. E. J. Bassey, and I am grateful for the valuable assistance of Mr. J. M. Irving and Mrs. A. Blecher, among many others.

REFERENCES

Bassey, E.J. (1978). Age, inactivity and some physiological responses to exercise. Gerontology, **24**, 66-77.

Bassey, E.J. & Fentem, P.H. (1980). Monitoring physical activity. In: W.A. Littler (ed.), Clinical and Ambulatory Monitoring, pp. 148-171. London: Chapman Hall.

Bassey, E.J., Bryant, J.C., Fentem, P.H., Macdonald, I.A. & Patrick, J.M. (1980). Customary physical activity in elderly men and women using long-term ambulatory monitoring of ECG and footfall. In: F. D. Stott (ed.), Proceedings of the 3rd International Symposium on Ambulatory Monitoring, 1979, pp. 425-432. London: Academic Press.

Bassey, E.J., Macdonald, I.A. & Patrick, J.M. (1982). Factors affecting heart-rate during self-paced walking. European Journal of Applied Physiology, **48**, 104-115.

Bassey, E.J., Patrick, J.M., Irving, J.M., Blecher, A. & Fentem, P.H. (1983). An unsupervised "aerobics" physical training programme in middle-aged factory workers: feasibility, validation and response. European Journal of Applied Physiology, **52**, 120-125.

Cunningham, D.A., Montoye, H.J., Metzner, H.L. & Keller, J.B. (1968). Active leisure activities as related to age among males in a total population. Journal of Gerontology, **23**, 551-556.

deVries, H.A. (1971). Exercise intensity threshold for improvement of cardio-vascular-respiratory function in older men. Geriatrics, **26**, 94-101.

Durnin, J.V.G.A., Blake, E.C. & Brockway, J.M. (1957). The energy expenditure and food intake of middle-aged Glasgow housewives and their adult daughters. British Journal of Nutrition, **11**, 85-94.

Durnin, J.V.G.A., Blake, E.C., Allan, M., Shaw, E.J. & Blair, S. (1961a). The food intake and energy expenditure of elderly women with varying-sized families. Journal of Nutrition, **75**, 73-76.

Durnin, J.V.G.A., Blake, E.C., Brockway, J.M. & Drury, E.A. (1961b). The food intake and energy expenditure of elderly women living alone. British Journal of Nutrition, **15**, 499-506.

Hodgson, J.L. & Buskirk, E.R. (1981). Role of exercise in aging. In: D. Danon (ed.), Aging: A Challenge to Science and Society, volume 1, pp. 189-196. Oxford: Oxford Medical Publications.

Ilmarinen, J. & Rutenfranz, J. (1979). Assessment of physical activity at work and during leisure time. In: F.D. Stott (ed.), Proceedings of the 2nd International Symposium on Ambulatory Monitoring, 1977, pp. 285-297. London: Academic Press.

Irving, J.M. & Patrick, J.M. (1982). The use of mechanical pedometers in the measurement of physical activity. In: F.D. Stott (ed.), Proceedings of the 4th International Symposium on Ambulatory Monitoring, 1981, pp. 369-376. London: Academic Press.

Katz, S., Branch, L.G., Branson, M.H., Papsidero, J.A., Beck, J.C. & Greer, D.S. (1983). Active life expectancy. New England Journal of Medicine, **309**, 1218-1224.

LaPorte, R.E., Cauley, J.A., Kinsey, C.M., Corbett, W., Robertson, R., Black-Sandler, R., Kuller, L.H. & Falkel, J. (1982). The epidemiology of physical activity in children, college students, middle-aged men, menopausal females and monkeys. Journal of Chronic Diseases, **35**, 787-795.

Paffenbarger, R.S., Wing, A.L. & Hyde, R.T. (1978). Physical activity as an index of heart attack risk in college alumni. American Journal of Epidemiology, **108**, 161-175.

Patrick, J.M. (1983). Infra-red monitoring. Ergonomics, **26**, 813-814.

Patrick, J.M., Bassey, E.J. & Fentem, P.H. (1982). Changes in body fat and muscle in manual workers at and after retirement. European Journal of Applied Physiology, **49**, 187-196.

Patrick, J.M., Bassey, E.J. & Fentem, P.H. (1983). The rising ventilatory cost of bicycle exercise in the seventh decade. Clinical Science, **65**, 521-526.

Patrick, J.M., Bassey, E.J., Irving, J.M., Blecher, A. & Fentem, P.H. (1985). Objective measurements of customary physical activity in elderly men

and women before and after retirement. Quarterly Journal of Experimental Physiology (in press).

Renfrew, J.W., Moore, A.M., Grady, C., Robertson-Tchabo, E.A., Colburn, T.R., Smith, B.M., Cutler, N.R. & Rapoport, S.I. (1984). A method for measuring arm movements in man under ambulatory conditions. Ergonomics, **27**, 651-662.

Shephard, R.J. (1978). Physical Activity and Aging. London: Croom Helm.

Sidney, K.H. (1981). Cardiovascular benefits of physical activity in the exercising aged. In: E.L. Smith & R.C. Serfass (eds.), Exercise and Aging, pp. 131-147. Hillside, NJ: Enslow.

Sidney, K.H. & Shephard, R.J. (1977). Activity patterns of elderly men and women. Journal of Gerontology, **32**, 25-32.

Stunkard, A. (1960). A method of studying physical activity in man. American Journal of Clinical Nutrition, **8**, 595-600.

EFFECTS OF AGEING ON HUMAN HOMEOSTASIS

K. J. COLLINS and A. N. EXTON-SMITH
Department of Geriatric Medicine,
University College School of Medicine, St. Pancras Hospital, London, U.K.

INTRODUCTION

From the evidence of a few longitudinal and many cross-sectional studies, human ageing can be characterised by a general decline in functional competence of the whole organism and a reduced capacity to respond as efficiently as younger individuals to internal and external stresses. This has become embodied in homeostatic theories of ageing, seen as a destabilisation of the internal environment of the organism and a limitation of the ability to bring about adaptive reactions. Because of their widespread influence on many integrated homeostatic mechanisms, the neural and endocrine centres of control in the hypothalamus are sometimes identified as possible centres of influence in the ageing process. Thus, theories attributing a key role to loss of sensitivity of hypothalamic neurones to neural or endocrine feedback signals have been proposed (Dilman, 1971; Dilman et al, 1979). Finch (1975, 1978, 1980) sees the loss of catecholamines in the hypothalamus and other parts of the brain as an important expression of the ageing process.

These theories seek to explain declining control of homeostasis in terms of metabolic and organisational deficiencies in hypothalamic nerve cells or their receptors. Though a number of observations testify to significant changes in centres of neural control which may be of value as biological markers of ageing (Reff & Schneider, 1982) it is probable that the decline in functional organisation of hypothalamic neurones controlling homeostasis is but an integral part of the overall ageing process, for structural and functional age changes can also be demonstrated in peripheral afferent and effector components of most physiological systems during ageing. The purpose of the following discussion is to provide a perspective to age-related changes in human homeostasis and, with the aid of one major system, that controlling body temperature, to show how "detraining" of physiological functions and sub-clinical pathological processes may influence control of homeostasis in old age.

LIMITATION OF HOMEOSTATIC ADJUSTMENTS

When applied other than in the broad sense originally suggested by W.B. Cannon as "the totality of steady states sustained by an organism", the concept of homeostasis may take on a number of different meanings. In the physiological context it describes the preservation of the constancy of internal cellular conditions, the sum of all physiological and biochemical regulations that maintain the steady state. Physiological homeostasis involves two interlinked processes, the maintenance of equilibrium in a system which possesses the property of inertia, and self-regulation by the organism which displays the property of adaptability. During morphogenesis, homeostasis refers to the process of development along a determined channel of growth, and in human genetics it implies the stabilisation of the gene pool representing the greatest degree of heterozygosity.

In human biology, homeostatic steady states are attained within the same limits in all races although the exact levels may be set at values depending on particular factors such as activity or environment. Although physiological equilibrium levels may not change during ageing there is evidence, some of which will be presented in this paper, of a change in inertia of the system.

Under resting conditions, there is no marked difference in intracellular composition between young and old individuals (e.g. Timiras, 1972), very little change with age in plasma composition and pH (Shock & Yiengst, 1950; Morgan, 1983), and vital extracellular equilibrium levels such as blood sugar, pH and osmotic pressure are regulated closely and maintained even into advanced old age. Nevertheless, though the evidence suggests no marked changes in intra- and extracellular constituents, age changes have been observed in a small number of blood constituents, e.g. cholesterol (see Bellamy (1985), this volume; Bailey (1985), this volume).

Any serious threat to the inertia of a homeostatic system calls into play compensatory adjustments which eventually stabilise the internal environment. In the elderly, however, there are demonstrable disorders of homeostasis and impairment in adapting to environmental stresses (Collins, 1985). For example, following an intravenous or oral glucose load, glucose tolerance is progressively impaired with ageing (Andres & Tobin, 1974). There appears to be little change in the fasting unstressed blood glucose level, and after a glucose load blood glucose rises equally in all age groups for the first 40 minutes. After 2 hours, although there are large variances at all ages, the elderly show a significantly higher blood glucose level. There are other examples of reduced efficiency in

regulating the biochemical *milieu interne*. Artificially induced acidosis or alkalosis causes changes in blood pH which are readily restored to normal in healthy young and elderly individuals by compensatory renal, respiratory and metabolic adjustments. Older people, however, respond more slowly to the challenge, and the duration of the adjustment process is extended. Similarly, loading an elderly individual with saline results in a more prolonged natriuresis relative to a young person (Epstein & Hollenberg, 1976). Thus, poor adaptability to chemical and physical stresses in the aged usually results in wider swings in the "hunting" pattern of adjustment and a lengthening of the period necessary to re-establish equilibrium.

Within the normal range of physiological tolerance compensatory processes are capable of maintaining function without serious disability. There is considerable functional reserve to protect the organism and in old age biological competence may be maintained by responses which are called upon more frequently than in younger individuals. Eventually, disability due to loss of intrinsic repair and immune mechanisms, detraining through disuse, loss of functional capacity and accumulating pathological changes may combine to cause breakdown of normal equilibrium levels and failure of homeostasis. In the early work of Simms (1942), haemorrhage was used as a cause of mortality in controlled animal studies. It was shown that a relatively small linear change in homeostasis resulted in a large logarithmic increase in age-specific mortality. Faulty control mechanisms were indicated because it was the chance of death following a particular volume of blood loss which increased with age rather than a logarithmic relation between the volume of blood loss and age. Similarly, Strehler and Mildvan (1960) interpreted the results of these studies as being due to a failure of the control systems to respond normally in aged animals and that there were inadequate reserves of energy.

DETRAINING

Most organs of the body are affected by morphological and functional involution during ageing which lead to a gradual decline in performance. Much of the deterioration in functional ability arises from decreasing cell numbers and loss of structural units (see Bellamy (1985), this volume), with the effects particularly marked if the cells are neurones. Partly it is also due to a diminishing blood supply to tissues. The hypertrophy of muscle fibres and an improvement in blood supply to skeletal muscle with repeated use in training is a well-known example of organ adaptation. Conversely the atrophy which occurs in muscle of bed-ridden patients illustrates the effects of disuse. In the

elderly there is often an increasingly negative attitude towards physical and mental exertion, brought about by loss of power and functional ability and eventually this leads to some degree of disuse atrophy.

If there is diminution in the capacity of homeostatic systems to adapt to stresses, the disengagement from active physical and mental exertion might be regarded as a naturally protective behavioural adaptation. Indeed some earlier concepts of ageing regarded disengagement as desirable for meeting the constraints of ageing and conserving body resources. In contrast with these concepts is the view that maintained high levels of activity improve well-being during human ageing, reflecting, among other things, a lower incidence of ill health (Palmore & Maddoc, 1977). To develop this argument further it would be logical to infer that an athlete might age more slowly and live longer than a more sedentary individual. At present, there is no good evidence to support the idea that sustained physical training by an ageing athlete will prolong his life span (Shephard, 1978). However, longitudinal studies carried out at Duke University, U.S.A., suggest that maintained levels of activity promote a more successful old age.

The effects of physical deconditioning (detraining) and loss of physiological capacity in the elderly have been discussed elsewhere in this volume (Patrick, 1985). Detraining which is often, though not an inevitable, consequence of ageing, arises through failure to maintain even the minimum of activity. This may become apparent in the inability to co-ordinate complex functions and is manifested by a raised threshold of response, inflexibility of reflex responses, and effector organ involution. As the range of activities is reduced so will the range of environmental stresses to which an individual is exposed. This could lead to a functional atrophy of unused systems, at first compensated when one system covers up the consequences of another failing, but eventually it may result in the loss of fine control of homeostasis.

PHYSIOLOGICAL AND PATHOLOGICAL PROCESSES IN AGEING

The need to discriminate between physiological and pathological changes has for long challenged gerontologists. Korenchevsky's group at Oxford in the 1960s attempted to make this distinction and concluded that old age was an abnormal, pathological syndrome, in which physiological processes of ageing were complicated by the degenerative diseases of old age (Korenchevsky, 1961). It was considered then that it was impossible to find a human being who aged physiologically and whose old age, life span and death were physiologically normal. However, it was thought possible to discriminate between physiological

and degenerative changes through the study of certain cardiovascular diseases such as cardiac atrophy and arteriosclerosis, and also other neuro-endocrine disease, and that it was "potentially possible" that in the distant future true physiological ageing of the whole organism might be observed. It is still premature to claim that this goal has been reached for it is very difficult to identify sub-clinical pathological processes, and the distinctions made between "normal" and pathological ageing are inconclusive. Some careful work has been undertaken on selected populations who are apparently free of disease where the effects of ageing are found to assume a different character from that revealed by cross-sectional studies in less selected groups. Two examples will serve to illustrate these points in relation to age-changes in adaptability.

One example referred to earlier is the response of elderly people to a glucose load. The basal level of blood glucose is similar in the elderly and young, but the rise in blood glucose concentration after a glucose load is much greater on average in the elderly. It has been shown (Denham, 1972), however, that the change in the average rise is due to a much greater change in a minority of the elderly than in others, and that this minority have diabetes mellitus. The increase in age-specific frequency of diabetes mellitus, it was suggested, is probably additional to a universal deterioration in glucose homeostasis in the elderly.

Cardiovascular performance during exercise has been measured in many studies, and it has been frequently observed that there is a decline in many indices of cardiac function (Brandfonbrener, Landowne & Shock, 1955; Julius, Antoon, Whitlock & Conway, 1967; Raven & Mitchell, 1980). Though there is little evidence that changes occur in intrinsic cardiac muscle function (e.g. Gerstenblith, Lakatta & Weisfeldt, 1976) alterations in cardiac performance with age have been explained on the basis of changes in β-adrenergic stimulation of heart rate, myocardial contractility and decrease in arteriolar tone (Lakatta, 1980). Reports from the Baltimore Longitudinal Study on Ageing in the U.S.A. have challenged some of the long-accepted findings on age decrements in physiological performance which classically demonstrate a linear decrement in resting cardiac index of about 30% from the age of 30 to 80 years (Shock, 1972). Whereas the earlier cross-sectional studies of randomly selected populations showed an age decrement in cardiac function, the longitudinal Baltimore study examined a population which was physically active and with neither overt nor occult coronary artery disease (Rodeheffer, Gerstenblith, Becker, Fleg, Weisfeldt & Lakatta, 1984). No significant age-associated decline in cardiac

output at rest and during exercise was found in the Baltimore population of "healthy" adults between the ages of 25 and 79 years. Compensatory physiological adaptations during exercise did however occur; a slower heart rate possibly due to less-efficient β-adrenergic stimulation was balanced by an increased cardiac dilatation and increased stroke volume. The results of these investigations argue for a diminution in the response of cardiovascular target organs to catecholamine stimulation during the process of ageing (see also Collins, 1983). Although it is pointed out by the Baltimore workers that the absence of adrenergic modulation could explain all the age-related changes in the haemodynamic response to exercise, judgment should be reserved until the completion of the longitudinal study. Whilst it may be inferred that in the absence of detraining and of occult coronary disease there appears to be little age-related decline in cardiac output, the suggestion that changes in autonomic nervous control may occur with age does not resolve the question as to whether such changes are physiological or pathological. As has been observed by Exton-Smith (1979) variations in the degree of involution and consequent decline in function may at times be so accentuated as to become pathological, and the result is a state which is obviously a departure from "normal" ageing.

THERMAL HOMEOSTASIS AND AGEING

Thermoregulation is a paradigm of the integrative actions of component systems directed towards the maintenance of a stable internal environment. Homeothermy was clearly an important event in mammalian evolution but the physiological mechanisms of thermoregulation differ widely in mammalian species. In some respects, the processes of thermoregulation in the human are highly specialised to the species. Thermogenesis derived from the metabolism of brown adipose tissue is restricted essentially to the human newborn, while non-shivering thermogenesis contributes little to homeothermy in the adult. The human sweat response is unique in the animal kingdom; in no species is there a parallel mechanism for dissipating large quantities of heat from the body surface such as that provided by the evaporation of eccrine sweat from the secretion of atrichial glands in the human skin. Behavioural thermoregulation also varies considerably from one species to another and only human beings possess the intellectual capacity to acquire clothing for insulation and to construct sophisticated micro-climates for thermal protection.

It has been recognised for some time that heat regulation is often disturbed in old animals. Verzar (1956, 1960) showed directly that young rats (2-6 months old) exposed for 1 hour at -2.5°C were able to maintain a steady body

temperature whereas in older animals the core temperature decreased by about $1^{o}C$; and in even older (24-26 months old) animals it decreased by about $3^{o}C$. In ambient temperatures between $-10^{o}C$ and $-14^{o}C$, the core temperature of many young animals did not decrease by more than $1^{o}C$, while the effect of this temperature environment on old animals was catastrophic, with many dying after a fall of core temperature by 11 or $12^{o}C$. Similarly, at high ambient temperatures ($38^{o}C$) old animals demonstrated an inability to contend with the heat.

Thermal lability in extreme climates is usually recognised in the elderly population by the incidence of hypothermia and heat stroke which occurs particularly in the older group (Collins et al, 1977; Ellis, 1972). Collins and Exton-Smith (1983) have shown this to be due to an inability of the intrinsic thermoregulatory system to maintain thermeostasis as well as to external factors of climate.

Central nervous control of thermeostasis

One feature of the central nervous control of thermoregulation is the well-marked circadian rhythm showing deep body temperature falling to a minimum during sleep at night and rising to a maximum in the daytime. The body temperature rhythm appears to be synchronised with the sleep-wake cycle, but it has been found that there is an inherent difference in the way in which these two rhythmic changes are generated by 'pacemakers' in the brain. In some cases where a person is deprived of cues which time these rhythms, the sleep-wake cycle operates free and with a different periodicity in relation to the body temperature rhythm. This can be induced by constant illumination instead of normal light-dark cycles. Studies on squirrel monkeys (Moore-Ede & Sulzman, 1981) revealed that when this desynchronisation occurs in a cold environment, there is an inability to maintain normal body temperature and the squirrel monkeys become hypothermic. In the human there is evidence that desynchronisation of circadian rhythms arises more frequently with increasing age (Wever, 1979) and this may represent a change in the inertia of centres controlling homeostatic systems resulting in increased lability of thermoregulation.

In studies of behavioural thermoregulation in groups of healthy young and old people, subjects were given control over the environmental temperature in the laboratory and asked to find the level of temperature that gave optimum thermal comfort. Although both the young and old groups chose the same

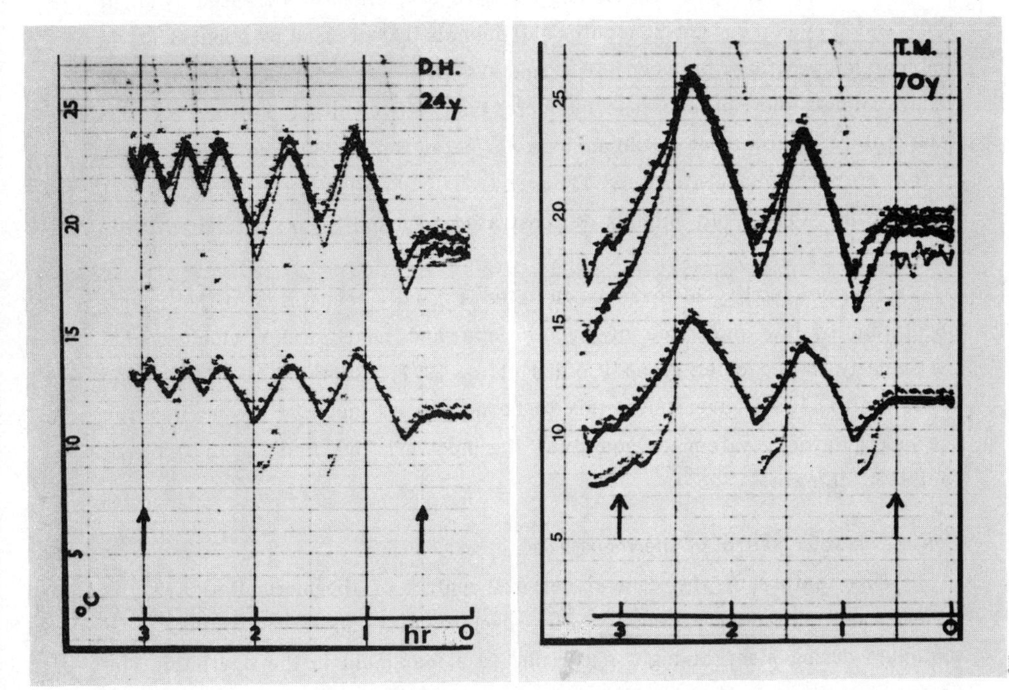

Figure 1: Behavioural thermoregulation by an elderly man aged 70 and a young man aged 24, individually controlling the temperature of a standard room. Room temperature was maintained at 19°C for 30 minutes before the remote-control period (time scale from right to left). Upper two traces record room temperature measured at 2 metres and at table height, the lower trace is wet-bulb temperature.

optimum temperature, the pattern of variation around this optimum was quite different. The young subjects allowed the temperature to vary by 3°C at the beginning but only by 1°C of the optimum after 3 hours (mean of 13 young subjects). The elderly showed the same 3°C deviations at the beginning but had increased the amplitude of oscillations to about 4.5°C after 3 hours (mean of 17 elderly of 70 years of age or more) (Collins et al, 1981). Figure 1 illustrates these differences in behavioural control of thermoregulation in a young and elderly subject. The physiological thermoregulatory responses appear to show the same differences in set-point control of body temperature but with the bandwidth increasing in old age and large displacements of the controlled variables being necessary to induce a corrective response in the elderly.

Changes in intrinsic components of the thermoregulatory system

Both sensory and motor responses to temperature change appear to become blunted with ageing. It is known that thermoreceptors in the skin of primates such as the ape are highly dependent for optimum function on a good oxygen supply (Iggo & Paintal, 1977) and in old age the vascular supply to skin tissues is reduced. Reduced blood supply and loss of nerve cells could both influence peripheral thermal perception. An increase in the threshold of cutaneous sensibility might also be due to changes in collagen and elastic tissues of the skin (Hall, 1976). On the whole the elderly perform much worse than the young in tests of ability to perceive temperature differences (Collins & Exton-Smith, 1981).

Effector systems of thermoregulation have received more study, and major changes have been described in shivering, sweating and vasomotor responses in the elderly. When groups of fit, active elderly and young adults, who have been taking no medication are compared, differences in these functions are often only slight. Investigation in our laboratory show equivalent increases in shivering thermogenesis in healthy 30-year-olds and in 70-year-olds undergoing standard convective cooling tests, though slight falls in temperature tend to occur more frequently in the elderly. Differences in thermogenesis exhibited by old and young groups selected for fitness are more subtle. High peaks of muscle contraction achieved by young people are not usually attained by the elderly and there is often a longer latent period required to produce maximum shivering. A contributory factor may be loss of motor power in the muscles themselves (Gutmann & Hanzlikova, 1976; Ermini, 1976) which may be due to a decrease in muscle bulk and to changes in distribution of muscle fibre types or impairment in the competence of the neuromuscular junction with age.

A number of investigations have shown abnormal vasoconstrictor patterns in elderly people exposed to cold (e.g. Wagner, Robinson & Marino, 1974; Collins et al, 1977). Some healthy elderly people, perhaps 20% of those over the age of 70, do not experience rapid vasoconstriction on cooling (Table 1). Furthermore, in longitudinal studies the proportion with poor constrictor responses to cooling increases during the course of an 8-year period (Table 1). In most young people it is possible to demonstrate transient bursts of vasoconstrictor activity occurring approximately 2-3 times a minute in neutral temperature conditions. This vasoconstrictor rhythm appears to originate in the central nervous system because electrical recording from sympathetic nerves supplying the blood vessels show a similar rhythm (Bini, Hagbarth, Hynnin & Wallin, 1980). In many elderly

Table 1. Mean hand blood flow in elderly and young subjects

	Young				Elderly			
	n	Age (yrs)	RBF	EC/EN%	n	Age (yrs)	RBF	EC/EN%
1972								
a.m.	22	26 (4)	13 (5)	30 (17)	74	73 (6)	11 (7)	40 (22) *
p.m.	23	26 (4)	13 (7)	27 (20)	55	73 (6)	15 (9)	39 (23) *
					21	72 (4)	11 (5)	34 (16)
1976	10	27 (7)	9 (4)	36 (18)	21	76 (4)	11 (4)	45 (22)
1980	13	23 (3)	15 (7)	26 (17)	21	80 (4)	12 (5)	53 (26) **
1983								
Insulators	12	23 (4)		16 (8)	8	73 (7)		41 (20) **
Shiverers	11			33 (19)	10			51 (25) *

EC/EN% young v. elderly:
** $p < 0.005$
* $p < 0.01$

Mean hand blood flow (ml/100 ml tissue/min ± SD) in elderly and young subjects exposed to cooling in air at 15°C measured by venous occlusion plethysmography. In 1983 blood flow was measured in air at 20°C by photoelectric pulsimeter. Resting blood flow (RBF) is measured and the degree of vasoconstriction expressed by blood flow at the end of cooling (EC) as a percentage of blood flow at the end of a neutral temperature period (EN). Twenty-one elderly subjects were included in a longitudinal study from 1972 to 1980. In 1983, subjects were divided into those who possessed a predominantly vasoconstrictor pattern (insulative) and those with a vasoconstriction plus shivering pattern of thermoregulation.

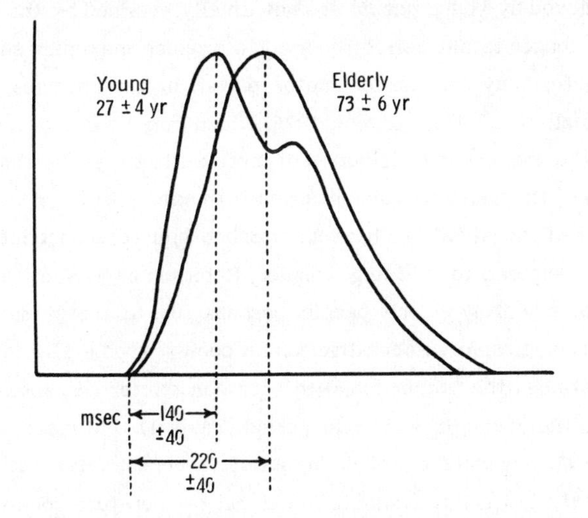

Figure 2. Pulse volume measured at the thumb by photoelectric plethysmograph in a neutral (23°C) environment. Mean values for time to pulse crest ± SD are shown for 16 young adult and 16 'healthy' elderly subjects.

people such a rhythmic vasomotor pattern is absent or difficult to detect (Collins et al, 1982; Collins, 1983) and this suggests altered sensitivity of the vasomotor control system.

In a recent study (Collins et al, 1985) mean deep body temperature fell significantly more in clothed, healthy elderly (means of 63-70 year olds) than in young adults when exposed for 2 or more hours to a cold environment of 6°C. In this climate, there was a marked rise in blood pressure which was significantly greater in the elderly than in the young. Onset of the blood pressure rise was, however, slower in the elderly but eventually a higher blood pressure (systolic and diastolic) was attained than in the young. These differences illustrate the characteristic longer latent period and impaired hunting pattern of homeostatic responses which are frequently seen in old age. There were no conspicuous improvements in thermogenesis (shivering responses) or insulative vasoconstriction after seven or more days repeated exposure to cold. Vasoconstriction was greater in the young both before and after cold acclimatisation.

Although such observations may indicate differences in peripheral vasomotor control by nervous structures, it is difficult to dismiss the possible effects of intrinsic pathological changes in the smooth muscle of the arterioles. One method of attempting to analyse this by non-invasive techniques is to measure the compliance of vessel walls (Ring et al, 1959). It has been demonstrated (Collins et al, unpublished) that there are significant differences in the contour analysis of pulse volumes in apparently healthy elderly people compared with young men (Fig. 2). The time to the pulse crest is increased in elderly subjects in neutral, cool and warm conditions, suggesting that compliance of the vessels has been reduced by atherosclerosis or by an increase in arterial wall stiffness. This is confirmed by the absence of a dicrotic wave in the elderly pulse volume traces (Fig. 2). In this particular group of 'healthy' elderly the degree of vasoconstriction on cooling was not significantly different from the younger group in spite of the fact that vascular compliance was demonstrably less in the elderly. Further, the change in compliance may be due to arterial stiffening (a 'natural' ageing process) and/or atherosclerosis (a 'disease' process) (Avolio et al, 1983). Clearly the sharp distinction often made between ageing and disease depends to some extent on the magnitude of change and may be an artificial construct.

CONCLUSIONS

The efficiency of homeostatic regulation declines in old age resulting in a reduced capacity to react to internal and external stresses and a limited ability

to adapt to new environments. Thermal homeostasis provides a good model with which to study these changes which are characterised by a loss of fine control and with wider swings in the hunting pattern of response. There is often inadequacy in thermogenesis and in the insulative vasoconstrictor reaction to cold stress and this may be partly due to disuse atrophy of organs and detraining of physiological systems. Effector responses show alteration in organ performance with diminished functional reserve and a longer latent period of response. It is, however, often difficult to make a clear distinction between pathological and physiological events which modify these changes. The assumption at present is that ageing represents a loss of cellular information which may involve processes of faulty transcription/translation and errors in macro-molecular synthesis. Such changes are likely to increase the accumulation of 'noise' in cellular copying and be expressed in the whole organism as reduced ability of homeostatic systems to bring about adaptive and stabilising reactions.

REFERENCES

Andres, R. & Tobin, J.D. (1974). Ageing and the disposition of glucose, In: V. Cristofalo, T. Roberts & R. Adelman (eds.), Advances in Experimental Medicine and Biology, vol. 61, pp. 239-249. New York: Plenum.

Avolio, A.P., Chen, S.-G., Wang, R.-P., Zhang, C.-L., Li, M.-F. & O'Rourke, M.F. (1983). Effects of aging on changing arterial compliance and left ventricular load in a Chinese urban community. Circulation, **68**, 50-58.

Bailey, A. (1985). Can we tell our age from our biochemistry? (This volume.)

Bellamy, D. (1985). Cell death and the loss of structural units of organs. (This volume.)

Bini, G., Hagbarth, K.E., Hynninen, P. & Wallin, B.G. (1980). Thermoregulatory and rhythm-generating mechanisms governing the sudomotor and vasoconstrictor outflow in human cutaneous nerves. Journal of Physiology, **306**, 537-552.

Brandfonbrener, M., Landowne, M. & Shock, N.W. (1955). Changes in cardiac output with age. Circulation, **12**: 557-566.

Collins, K.J. (1983). Autonomic failure and the elderly. In: R. Bannister (ed.), Autonomic Failure, pp. 489-507. Oxford: Oxford University Press.

Collins, K.J. (1985). Disorders of homeostasis, In: A.N. Exton-Smith & M. Weksler (eds.), Practical Geriatric Medicine, pp. 74-83.

Collins, K.J. & Exton-Smith, A.N. (1981). Urban hypothermia, thermal perception and thermal comfort. In: J. M. Adam (ed.), Hypothermia Ashore and Afloat, pp. 158-176. Aberdeen: Aberdeen University Press.

Collins, K.J. & Exton-Smith, A.N. (1983). Thermal homeostasis in old age. Journal of the American Geriatrics Society, **31**, 519-524.

Collins, K.J., Doré, C., Exton-Smith, A.N., Fox, R.H., Macdonald, I.C. & Woodward, P.M. (1977). Accidental hypothermia and impaired temperature homeostasis in the elderly. British Medical Journal, **1**, 353-356.

Collins, K.J., Exton-Smith, A.N. & Doré, C. (1981). Urban hypothermia: preferred temperature and thermal perception in old age. British Medical Journal, **282**, 175-177.

Collins, K.J., Easton, J.C. & Exton-Smith, A.N. (1982). The ageing nervous system: impairment of thermoregulation. In: M. Sarner (ed.), Advanced Medicine, pp. 250-257. Bath: Pitman.

Collins, K.J., Easton, J.C., Belfield-Smith, H., Exton-Smith, A.N. & Pluck, R. (1985). Effects of age on body temperature and blood pressure in cold environments. Clinical Science, **69**, 465-470.

Denham, M.J. (1972). The value of random blood glucose determinations as a screening method for detecting diabetes mellitus in the ageing patient. Age and Ageing, **1**, 55-60.

Dilman, V.M. (1971). Age-associated elevation of hypothalamic threshold to feedback control, and its role in development, ageing and disease. Lancet **1**, 1211-1219.

Dilman, V.M., Ostroumova, M.N. & Tsyrlina, E.V. (1979). Hypothalamic mechanisms of ageing and of specific age pathology. II. On the sensitivity threshold of hypothalamic-pituitary complex to homeostatic stimuli in adaptive homeostasis. Experimental Gerontology, **14**, 175-181.

Ellis, F.P. (1972). Mortality from heat-illness and heat-aggravated illness in the United States. Environmental Research, **5**, 1-58.

Epstein, M. & Hollenberg, N.K. (1976). Age as a determinant of renal sodium conservation in normal man. Journal of Laboratory and Clinical Medicine, **87**, 411-417.

Ermini, M. (1976). Ageing changes in mammalian skeletal muscle. Gerontology, **22**, 301-316.

Exton-Smith, A.N. (1979). Special features of diseases in old age, In: Geriatrics. Guidelines in Medicine, vol. I, pp. 17-34. Lancaster: MTP Press.

Finch, C.E. (1975). Ageing and the regulation of hormones. In: V. Cristofalo, T. Roberts & R. Adelman (eds.), Advances in Experimental Medicine and Biology, p. 229. New York: Plenum.

Finch, C.E. (1978). The brain and ageing. In: J.A. Behnke, C.E. Finch & G.B. Moment (eds.), The Biology of Ageing, pp. 301-309. New York: Plenum.

Finch, C.E. (1980). Neural and endocrine mechanisms in aging. In: R.T. Schimke (ed.), Biological Mechanisms in Aging. U.S. Department of Health & Human Services, NIH Publication No. 81-2194, pp. 536-701. Bethesda, Maryland.

Gerstenblith, G., Lakatta, E.G. & Weisfeldt, M.L. (1976). Age changes in myocardial function and exercise response. Progress in Cardiovascular Diseases, **19**, 1-21.

Gutmann, E. & Hanzlikova, V. (1976). Fast and slow motor units in ageing. Gerontology, **22**, 280-300.

Hall, D.A. (1976). Ageing of Connective Tissue. London: Academic Press.

Hugin, F. & Verzar, F. (1956). Versagen der Warmeregulation bei Kalte als Alterserscheinung Symp. uber exp. Alternsforschg., p. 96. Birkhauser: Verlag.

Iggo, A. & Paintal, A.S. (1977). The metabolic dependence of primate cutaneous cold receptors. Journal of Physiology, **272**, 40P.

Julius, S., Antoon, A., Whitlock, L.S. & Conway, J. (1967). Influence of age on the haemodynamic response to exercise. Circulation, **36**, 222-230.

Korenchevsky, V. (1961). Physiological and pathological processes of ageing, In: V. Korenchevsky & G.H. Bourne (eds.), Physiological and Pathological Ageing, pp. 1-4. Basel: Karger.

Lakatta, E.G. (1980). Age-related alterations in the cardiovascular response to adrenergic mediated stress. Federation Proceedings, **39**, 3173-3177.

Moore-Ede, M.C. & Sulzman, F.M. (1981). Internal temporal order, In: J. Aschoff (ed.), Handbook of Behaviour Neurobiology, pp. 215-241. New York: Plenum.

Morgan, D.B. (1983). The impact of ageing - present and future. Annals of Clinical Biochemistry, **20**, 257-261.

Palmore, E. & Maddox, G.L. (1977). Sociological aspects of ageing, In: E.W. Burse & E. Pfeiffer (eds), Behaviour and Adaptation in Late Life, 2nd edn. Boston: Little, Brown & Co.

Patrick, J.M. (1985). Customary physical activity in the elderly. (This volume.)

Raven, P.B. & Mitchell, J. (1980). The effect of aging on the cardiovascular response to dynamic and static exercise, In: M.L. Weisfeldt (ed.), The Ageing Heart, pp. 269-296. New York: Raven Press.

Reff, M.E. & Schneider, E.L. (1982). Biological markers of aging. U.S. NIH Publication No. 82-2221. Bethesda, Maryland.

Ring, G.C., Kurbatov, T.C. & Shannon, C.J. (1959). Changes in central pulse and finger plethysmography during ageing. Journal of Gerontology, **14**, 189-191.

Rodeheffer, R.J., Gerstenblith, G., Becker, L.C., Fleg, J.L., Weisfeldt, M.L. & Lakatta, E.G. (1984). Exercise cardiac output is maintained with advancing age in healthy human subjects: cardiac dilatation and increased stroke volume compensate for a diminished heart rate. Circulation, **69**, 203-213.

Shephard, R.J. (1978). Physical Activity and Ageing. London: Croom-Helm.

Shock, N.W. (1972). Energy metabolism, calorie intake and physical activity of the aged, In: L.A. Parlson (ed.), Nutrition in Old Age, pp. 12-23. Symposium of the Swedish Nutrition Foundation. Uppsala: Almqvist & Wiksell.

Shock, N.W. & Yiengst, M.J. (1950). Age changes in acid-base equilibrium of the blood of males. Journal of Gerontology, **5**, 1-4.

Simms, H.S. (1942). The use of measurable cause of death (haemorrhage) for the evaluation of ageing. Journal of General Physiology, **26**, 169-178.

Strehler, B.L. & Mildvan, A.S. (1960). General theory of mortality and ageing. Science, **132**, 14-21.

Timiras, P.S. (1972). Developmental Physiology and Ageing, pp. 542-563. London/New York: MacMillan.

Verzar, F. (1960). Adaptation to environmental changes at different ages, In: B.L. Strehler, J.D. Ebert, H.B. Glass & N.W. Shock (eds.), The Biology of Ageing, pp. 324-327. Washington DC: American Institute of Biological Sciences.

Wagner, J.A., Robinson, S. & Marino, R.P. (1974). Age and temperature regulation of humans in neutral and cold environments. Journal of Applied Physiology, **37**, 562-565.

Wever, R.A. (1979). The Circadian System of Man: Results of Experiments under Temporal Isolation. New York: Springer-Verlag.

INDEX

PUBLISHED SYMPOSIA OF THE
SOCIETY FOR THE STUDY OF HUMAN BIOLOGY

1 The Scope of Physical Anthropology and its Place in Academic Studies Ed. D.F. ROBERTS and J.S. WEINER
2 Natural Selection in Human Populations Ed. D.F. ROBERTS and G.A. HARRISON
3 Human Growth Ed. J.M. TANNER
4 Genetical Variation in Human Populations Ed. G.A. HARRISON
5 Dental Anthropology Ed. D.R. BROTHWELL
6 Teaching and Research in Human Biology Ed. G.A. HARRISON
7 Human Body Composition, Approaches and Applications Ed. J. BROZEK
8 Skeletal Biology of Earlier Human Populations Ed. D.R. BROTHWELL
9 Human Ecology in the Tropics Ed. J.P. GARLICK and R.W.J. KEAY
10 Biological Aspects of Demography Ed. W. BRASS
11 Human Evolution Ed. M.H. DAY
12 Genetic Variation in Britain Ed. D.F. ROBERTS and E. SUNDERLAND
13 Human Variation and Natural Selection Ed. D.F. ROBERTS (Penrose Memorial Volume reprint
14 Chromosome Variation in Human Evolution Ed. A.J. BOYCE
15 Biology of Human Foetal Growth Ed. D.F. ROBERTS
16 Human Ecology in the Tropics Ed. J.P. GARLICK and R.W.J. KEAY
17 Physiological Variation and its Genetic Basis Ed. J.S. WEINER
18 Human Behaviour and Adaptation Ed. N.G. BLURTON JONES and V. REYNOLDS
19 Demographic Patterns in Developed Societies Ed. R.W. HIORNS
20 Disease and Urbanisation Ed. E.J. CLEGG and J.P. GARLICK
21 Aspects of Human Evolution Ed. C.B. STRINGER
22 Energy and Effort Ed. G.A. HARRISON
23 Migration and Mobility Ed. A.J. BOYCE
24 Sexual Dimorphism Ed. F. NEWCOMBE *et al.*
25 The Biology of Human Ageing Ed. A.H. BITTLES and K.J. COLLINS

Numbers 1–9 were published by Pergamon Press, Headington Hill Hall, Headington, Oxford OX3 0BY. Numbers 10–24 were published by Taylor & Francis Ltd, 10–14 Macklin Street, London WC2B 5NF. Further details and prices of back-list numbers are available from the Secretary of the Society for the Study of Human Biology.

SOCIETY FOR THE STUDY OF HUMAN BIOLOGY

SYMPOSIUM SERIES: 25

The biology of human ageing

W9-BWQ-777